Abdulrahman Obaid Al-Youbi ·
Adnan Hamza Mohammad Zahed ·
Mahmoud Nadim Nahas ·
Ahmad Abousree Hegazy

The Leading World's Most Innovative Universities

 Springer

Abdulrahman Obaid Al-Youbi
King Abdulaziz University (KAU)
Jeddah, Saudi Arabia

Mahmoud Nadim Nahas
King Abdulaziz University (KAU)
Jeddah, Saudi Arabia

Adnan Hamza Mohammad Zahed
International Advisory Board of KAU
King Abdulaziz University (KAU)
Jeddah, Saudi Arabia

Ahmad Abousree Hegazy
King Abdulaziz University (KAU)
Jeddah, Saudi Arabia

This Springer imprint is published by the registered company Springer Nature Switzerland AG
The registered company address is: Gewerbestrasse 11, 6330 Cham, Switzerland

Acknowledgements

The authors express their thanks and appreciation to King Abdulaziz University for giving them the opportunity to compose this book. They extend their thanks to all who participated in completing this book.

The authors also thanks all members who contributed their efforts and ideas and all who duly cared and were keen for this work's completion, hoping that it would contribute to the progress of King Abdulaziz University, in particular, and Saudi universities in general.

The authors would also like to thank Mr. Tarek Youssef Sida for editing and proofreading the full text and translation.

Introduction

Innovation means the new ways or methods that differ from conventional methods used in creating or developing things and ideas. It is about processes expressing fundamental changes in thinking; production or products; processes or ways and methods of performance; or administrative organizations and structures. Other terminology, such as creativity and invention, might overlap with this term. The main objective behind innovation is most likely the positive change and significantly making things, ideas, or methods better than what they originally were before becoming innovative.

In many scientific and professional fields, innovation leads to increased productivity and mainly contributes to the development of organizational and national resources. Individuals described as "innovative" are most likely pioneers at work or field of specialization. This is also applicable to pioneering organizations.

Innovation is produced through exerting time and effort in researching into an idea, developing ideas, or marketing them for beneficiaries. All innovations start with creative ideas, where innovation makes significant changes in the product. Thus, innovation is the successful application of creative ideas in any institution, organization, or facility. Accordingly, creativity is the onset of innovation. Creativity is essential for innovation; however, it is insufficient on its own. Ideas should be practically inspected and experimented in real life to identify their effectiveness and associated processes, along with the management of such processes at minimal cost and less effort. If creative ideas are produced by individuals, innovation is produced by institutions and organizations incubating such creative ideas. Creativity is neither an inherited trait nor a genetic mutation, but a product of an environment that believes in its importance, upholds it, and supports creative initiatives. This is not only limited to college and postgraduate stage, but also starts much earlier. It is not an exaggeration to say that this should be started at primary school. Since the 1960s, a competition known as the Young Scientist has been held in the UK. In this competition, creative students from primary, middle, and high schools compete with research submission. This is given all kinds of support by their family, school, and local authority, as winning such a competition is considered an honor for the student, his family, his school, and the city that supports him.

Top educational institutions keep continuous development to achieve distinctive results. Some world universities excelled in terms of innovation, so they are reckoned and listed by Reuters, in association with Clarivate Analytics Company, as World's Most Innovative Universities.

In the light of the above, this book is introduced to present a summary of the factors affecting universities' ranking in World's Innovative Universities record, thus understanding how Arab universities can be listed in this record.

In particular, this book aims at achieving the following:

1. Identify the meaning of creativity and innovation,
2. Identify the methodology of ranking the World's Most Innovative Universities,
3. Identify the world outstanding universities and their global ranking in terms of innovation,

4. Identifying the European outstanding universities and their ranking among European universities in terms of innovation,
5. Identify the Asian-Pacific outstanding universities and their ranking among Asia-Pacific universities in terms of innovation,
6. Study of the world's top twenty outstanding universities in terms of innovation,
7. Study of some other world's outstanding universities in terms of innovation,
8. Study of some of the factors that help universities be innovative,
9. Identify how any Arab university can reach the level of the World's Most Innovative Universities.

Studying the development of leading universities in the field of education, learning, research, and innovation is a good start for excellence. Any university should shape its own way by itself instead of simulating other universities and make use of its professors, students, and infrastructure. In fact, a face-to-face meeting between professors and their students allows them to perform outstanding jobs and helps them create and innovate, enabling the university to perform vital functions including acquisition of new knowledge, making use of the previous findings and leading the new generation. A university that adapts itself to the new competitive environment depending on its points of strength and adopting new techniques will have a brilliant future since adaptation to what is new is the characteristic of leading universities.

Contents

About the Authors

Prof. Abdulrahman Obaid Al-Youbi has been the President of King Abdulaziz University (KAU) since 2016 and a Professor of Chemistry at KAU since 2000. He earned a Ph.D. in Physical Chemistry from Essex University, UK, in 1986. He has also been the President of the International Advisory Board (IAB) of KAU since 2015. Throughout his career, Prof. Al-Youbi has been an active researcher in his specialization, a passionate teacher, and an academic administrator. He has participated in many research projects and has published more than 150 papers in ranked scientific journals. He has also supervised many graduate students. He has held a variety of Academic Administrative positions at KAU including Chairman of the Chemistry Department, Vice-Dean of the Faculty of Science (1992–1999), Dean of the Faculty of Science (1999–2002), Vice President (2002–2009), and Vice President for Academic Affairs (2009–2016). In 2015–2016, he was Acting President of both KAU and Jeddah University. As President of KAU, Prof. Al-Youbi has devoted his position to strengthening excellence in academics and research with a dedication to developing an innovative culture. Through President Al-Youbi's leadership, KAU has remained the top university, not only in the Kingdom of Saudi Arabia, but also in the Arab World. His current focus is on expanding KAU's lead by continuing to build on its long-standing strengths in education, research, entrepreneurship, and Community Service to the people of the Kingdom of Saudi Arabia.

President Al-Youbi has participated in more than eighty committees, boards, teams, and working groups at the university level as well as at the Ministry of Education level. In particular, he has participated in the committees that have established new universities in the Kingdom, namely Taiba University, Jazan University, Tabuk University, and the Northern Border University. He has also attended many scientific conferences in the Kingdom and abroad.

Prof. Adnan Hamza Mohammad Zahed has worked as Consultant to the President of King Abdulaziz University (KAU) since 2016 and the Secretary-General of the International Advisory Board (IAB) of KAU since 2010. He was the KAU Vice President for Graduate Studies and Scientific Research (2009–2016) and worked before that as Dean of Graduate Studies (2007–2009), and before that, he was Vice-Dean in the Faculty of Engineering (1997–2007). Prof. Zahed has been a full professor in the Chemical Engineering Department at KAU since 1996. He has also worked in industry as General Supervisor (Consultant) in Saudi Badrah Company (Jeddah, KSA, 1995–1996), Deputy General Manager at Savola Food Company in Jeddah (1993–1995), and Deputy CEO of Tasali Company (Jeddah). He holds a B.Sc. in Chemical Engineering from King Fahd University of Petroleum and Minerals, KSA (1976), and a M.S. and Ph.D. in Chemical Engineering from the University of California (Davis), USA (1982). He has published six books, one patent, and more than 60 papers in international conferences and refereed journals, in addition to more than 75 technical reports written for the bodies who funded his projects. He has also been a co-author of several University Guides such as the Graduate Studies Guide, Applicable Theses Guide, Thesis Writing Guide, Graduate Studies Procedure Guide, Faculty of Engineering Prospectus, and Annual Report of Research Activities in the Faculty of Engineering. Prof. Zahed was included in Marquis Who's Who in the World 2006. He has participated in more than 80 committees at departmental, faculty, and university levels at KAU. In addition, he has participated in four academic accreditation meetings in the USA and in more than 25 local and international conferences, symposia, and forums. Prof. Zahed has visited a number of American universities as a delegate of the Saudi Ministry of Education.

Prof. Mahmoud Nadim Nahas worked as Consultant to the Vice President of Graduate Studies and Scientific Research of King Abdulaziz University (KAU) for about ten years. He then became the consultant of the Secretary-General of the International Advisory Board (IAB) of KAU. Prof. Nahas is a full professor in the Faculty of Engineering at KAU since 1996. He holds a B.Sc. in Mechanical Engineering from Aleppo University, Syria (1973), and an M.Sc. (1978) and a Ph.D. in Aeronautical Engineering from Cranfield University, UK (1981). He has published more than twenty books, one patent, and more than 80 papers in international conferences and refereed journals, in addition to more than 35 classified technical reports. He taught many engineering courses and developed many curricula and study programs. Nahas is also fond of simplifying scientific issues to the public. He has published hundreds of articles in newspapers and magazines in this regard. He is a regular writer in two magazines.

Prof. Ahmad Abousree Hegazy is a Professor of Mechanical Engineering at the Faculty of Engineering, Helwan University (Egypt) since 2001 and at King Abdulaziz University (KAU) since 2008. He earned a Ph.D. in Mechanical Engineering from Leeds University, UK, in 1985. He has served as the consultant of the Secretary-General of the International Advisory Board (IAB) of KAU since 2010. He was a consultant of KAU Vice Presidency for Graduate Studies and Scientific Research (2009–2016).

Prof. Hegazy earned a M.Sc. from Helwan University (HU) in Mechanical Engineering (1981). He was appointed Assistant Professor (1986) at HU, Associate Professor and later as a full professor of Mechanical Engineering (2001). He held some Academic Administrative posts such as Chairman of the Production Engineering Department and Vice-Dean of the Faculty of Engineering (HU).

Throughout his career, he has been an active researcher in powder metallurgy and engineering materials, a passionate teacher, and an academic administrator. He has authored more than 40 scientific and academic publications and supervised over 10 graduate students.

Prof. Hegazy was awarded the Overseas Research Students from Committee of Vice Chancellors and Principals of the Universities of the United Kingdom (1983–1985), a British Council award, UK (1987), and a Grant by the Foreign Relation Coordination Unit of the Supreme Council of Universities (Egypt) as a member of research teamwork in project titled (UEEIEMP).

1.1 Introduction

Since the last decades of the twentieth century, most countries, including the Kingdom of Saudi Arabia, have been seeking to transform their economies from depending on primary natural resources to a knowledge-based one, providing a permanent source for economic growth, and thus achieving sustainable development, and helping in diversifying the national economic framework, and multiplying income and wealth resources of the country other than primary resources, mainly crude oil.

In all countries of the world, sustainable economic development is facing challenges imposed by modern developments caused by globalization and fierce competition. Consequently, economic development now is in need for creativity and innovation, rising competitiveness for advanced industries, increasing its efficiency, and managing investment risks, in addition to its need for high-skilled human resources inputs. This is where universities' role comes into action as a basic resource with their human assets and research institutions capable of producing knowledge, creativity, and innovation represented in researchers, research centers, and scientific departments. Universities are considered a resource of economic development and the most important inputs of production processes for the knowledge-based economy in this century. Pioneering universities represent an integral part of the production system of innovations and technologies and its transferring to business and society [1].

As known, this is the age of information and knowledge, and this is due to the reliance on scientific research, creativity, and innovation. Also, the foundations of today are laid upon the achievements of the industrial revolution followed by information and communication revolution.

Information and communication revolution, supported by increased efforts and achievements of scientific research, lead to multiplied and varied knowledge of science and technology. Furthermore, the global openness and substantial development in the means of mobility and communication showed that the reached innovations, achievements, and inventions are not the end but still to be transformed and translated into the industrial practical field, achieving record profits and returns. This caused the transformation into economics of knowledge as a natural outcome of the effective transformation to knowledge societies in many countries.

Every country is keen on exploring natural resources and developing its means and capabilities. Recent studies show that the richest untapped resources in the Arab region, especially within the last two centuries, would be the power of the human mind. Although major industrialized countries have far preceded us toward technology renaissance since more than a century, the modern thinking renaissance march has not started until the late decades of the twentieth century, thus giving hope in catching up, provided that an initiative should be taken to embark into this field with the needed persistence, vision, and strategy.

Thinking is the main function distinguishing man than other creatures. It is a valuable and continuous mental process for achieving a certain goal, may be represented in a solution for a problem, or generating an effective idea to reach the main goal which is moving to a better condition.

One who studies or considers world countries would find out that most points of comparison center around what is called: Developed and developing countries; rich and poor countries; first-world and third-world countries. Some researchers focus on certain criteria, such as individual income level, availability of resources, industry, and technical development. These criteria actually represent a part of the needed requirements for transforming communities or countries from poor into rich countries, or from developing into advanced countries on the right path. The key is the countries' ability to narrow the knowledge gap, without which third-world countries or countries aiming at transformation into communities and economies of knowledge would be unable to transcend to the level of developed countries [1].

The Leading Worldâ s Most Innovative Universities,
https://doi.org/10.1007/978-3-030-59694-1_1

Accordingly, narrowing or filling the knowledge gap is the doorstep to moving from regress to progress, whether it is economic, scientific, or cultural regress in general. It should be taken into consideration that nations' wealth is no longer centered on natural material and sources of wealth such as natural resources and lands, and it is rather represented in knowledge. That explains how countries that lack natural resources, such as Japan, Switzerland, Denmark, and Singapore, turned wealthy because of their knowledge resources. Such countries have become among the world's richest and highest GDP per capita countries. Meanwhile, some of the richest countries in terms of natural resources, such as Russia and Brazil, are of low GDP per capita income in comparison with other major countries [1].

Albeit thinking is individual at first place, its role gets more important in different institutions and countries on an individual, collective, or institutional scale, where thinking becomes a vital element adding to value, which leads to achieving excellence and superiority, and promotes leadership and competitiveness. Knowledge and thinking are the main tools for creativity for their contribution to continuous development of institutions and countries. Knowledge is an intellectual capital representing the thought and action outputs of selected operatives endowed with knowledge and organizational abilities to produce and develop new ideas, and learn continuously how to improve their institutions for optimizing their competitive capabilities. Effective institutions, and even countries that follow progress path primarily in the age of knowledge, are based upon the effort, creations, and innovations of human resources and intellectual capital.

1.2 Creativity and Innovation

Creativity has many definitions such as the ability to produce a work that is of two qualities: be unprecedented and adequate [2]; or it is the ability to see potential things that are not realized by others [3]. It is also defined as a sensitive and critical process that results in the generation of new ideas [4].

As for innovation, it may be perceived as applying a new product or an innovatively developed product, whether it is a commodity, a service, an application of production processes, new methods of marketing, or modern organizational methods. On another hand, innovation is defined as intended incorporation and application of creative ideas in a business, or a system of businesses, including processes, products, and new procedures in business leading to creating something of value that may be accepted and marketed in society. Innovation can be also perceived as new ideas produced and promoted in markets for the first time, ensuring that something new is presented at the market [5].

1.3 Relationship Between Creativity and Innovation

From the definitions provided, we can see the overlap between creativity and innovation, but it is safe to say that creativity is a permanent resource for innovation, while innovation is an application and implementation of creativity; thus, creativity and innovation are inseparably related, which reflects their complementarity in providing what is new and adding value.

There are many theories about creativity and innovation, but here, we highlight the concerned theory in terms of universities' role in building the economy of knowledge, which mainly focuses on creativity and innovation. This is because creativity gives no space for innovation unless it finds its way to application and marketing. A creative idea is nothing more than a realization or a vision; then, it is transferred into a product and becomes an innovation.

The end-product theory (or approach) tackles this interrelationship. The theory refers to the interrelationship between creativity and innovation, where the chain starts with creativity as a new applicable idea and then ends in the form of a product, a business, or some output. The creative experiment is the opposite of the reproduction experiment [6].

1.4 Importance of Knowledge for Creativity and Innovation

Scientific research, basic theoretical or practical studies, is extremely important in achieving the utmost benefit from scientific knowledge in innovation and development. Knowledge has become the focus of attention of countries that realized its importance in achieving sustainable development and global companies seeking to maintain the stature of their product around the world, in addition to scientific institutions (universities and research centers).

Development processes in developed countries have become to revolve around the concept of knowledge-based economy, as it leads to creativity and innovation. A society with knowledge is one that is capable of dealing, interacting, contributing, participating, producing, creating, and innovating.

The importance and role of knowledge in creativity and innovation can be summarized in the following points [7]:

1. Using and employing knowledge in fields of business and services; commodity production; and all economic activities,

2. Scientific and practical knowledge is the key basis for achieving innovations, discoveries, and technological inventions,
3. Continuous increase in investments directly related to knowledge, which results in forming knowledge capital represented by Intangible assets,
4. Continuous increase in institutions and projects working in terms of producing and using knowledge, represented by information, communication and software companies, and research and advisory institutions,
5. Preparing opportunities for institutions to focus on the most innovative departments and to encourage individuals of these departments to be continuously creative and innovative.

1.5   Creativity and Innovation in Universities

Universities have become the source of power in the knowledge-based economy of the century, as they represent an integral part of the production chain of innovations and skills and transferring them to business and society.

Many universities have a lot of programs, curricula, workshops, and mechanisms that have been developed to encourage students on creativity and innovation and improve their cognitive functions. Cropley et al. consider that most creativity mechanisms and programs, such as brainstorming and other mechanisms and methods of producing creative ideas, address only one side of creativity sources [8].

Consequently, innovation is a main factor that plays an important role in universities' tasks and is represented in using and employing the activities and outputs of universities' mission in education and scientific research fields, employing them in community service and thus achieving revenue for universities. As for knowledge triangle, including education, scientific research, and innovation, we denote that innovation is the resultant of education and scientific research through creative knowledge and reaching new ideas, business, and products, or developing and improving what is already there, transferring knowledge beyond universities, participating with business and society, and applying achieved innovations and transferring them to commodities and services. At the end of the twentieth century, world universities in developed countries started offering study programs for creativity, innovation, and entrepreneurship and focusing on applied researches solving business problems in addition to paying attention to knowledge transfer and exchanging activities between universities and business institutions. Furthermore, they focused on intermediary technology institutions, technology centers, technology transfer firms, business incubators, start-ups establishing projects for industrial investment of knowledge

and the development of knowledge transfer partnership. This aimed at developing and achieving a business income from the outputs of the scientific research activities and marketing them in business sector to compensate the decline in public funds for universities.

Governments of these countries encouraged such role out of urging universities to transfer and transform the outcomes of science, technology, creativity, and innovation to the service of economy. As a means of supporting universities role in creativity and innovation and building a knowledge-based economy, governments established a new type of organizational structures for boosting creativity and innovation processes and their transfer from universities to business institutions, and developing what is known as knowledge mediators and knowledge brokerage, represented in institutions whose job is to facilitate knowledge transfer, use, and sharing between universities and business sector through a two-way process. One way is having a university more open to community and the world, while the other is building a strong relationship between community and university, where they work on providing expertise, marketing knowledge, increasing academics, and executives focus on mutual interests, publishing, and implementing innovations for achieving their main objective, which is supporting and developing business, thus achieving economic prosperity through utilizing the university's capabilities for providing high-efficiency skills and high-quality basic, practical researches and for spreading the research and innovation culture.

The true value of any institution, including universities, lies in creative capital or intellectual human capital which means the mental capability for generating new adequate ideas with high quality. Creative intellectuals would be capable of employing their ideas by transforming them into valuable products and services achieving leadership, securing profits and good competitive position for their institution. Creative capital or intellectual human capital is one of the most important resources through which institutions can be competitive, representing the origin and basic resource for knowledge [9]. This capital, as a basic source for creativity and innovation, represents a strategic resource with high knowledge, mental energy, and distinguished skills and capabilities.

Universities are the institutions that embrace intellectual human capital the most. They serve as knowledge institutions that comprise teaching staff, researchers, and students, thus being a huge stock of skills and potentials that can bring constantly growing returns of creativity, generation of ideas, and innovation in an organizational environment and culture. The objective of universities' executive administration is centered on the proper management and utilization of the intellectual human capital. This represents a challenge for the university administration, but the greatest challenge yet is to transform ideas into intellectual or material values for work, production, and life; in other words, into knowledge in general. This indicates the transformation from thinking and creativity into innovation; it also suggests the transformation from creative university stage into innovative productive university stage. Management of intellectual human capital has become a top priority for universities for achieving many goals, such as creativity, social research production, applied scientific research publishing and marketing, and bringing about welfare to their communities [10].

Innovative universities represent an essential basis for delivering and exchanging knowledge, information, and human resources industry, making them either a strength or a weak factor of communities. Therefore, a university can be defined as the "engine of progress, beacon of enlightenment and forward-looking intellectual pillar." At the same time, it is a factory for preparing, forming, and qualifying successive generations and an entity entrusted with solving community problems through practicing them, creatively interact with them and objectively understand their dimensions [11]. Universities have to activate participation, transfer and exchange of information, and transform such information in individuals' minds to achieve required competitiveness; in order to be based on learning and knowledge generation; or obtaining, transferring, or utilizing information to reach new knowledge for solving their problems; or commercially investing it in cooperation with other different sectors of community.

Universities emerged as primary drivers in economics of knowledge, and they are expected to play vital roles in innovation and technology development. While a leading research university has become a key requirement and an important asset for economy, it is insufficient in itself for creating strong regional economic growth due to universities' tendency toward being catalysts for technology innovation, more than being drivers. Hence, universities must make use of their various strength points in boosting creativity and innovation processes and play more roles and exert further efforts in developing products and generating start-ups for applying achieved innovations, increasing graduate employment and effectively contributing in increasing their financial resources, thus contributing in economic development for community, as shown in Fig. 1.1.

1.6   The Role of Innovation in Development

Knowledge economy is based on knowledge and innovation and their consequent advanced techniques and regulations granting national economy high capacities of competitiveness and sustainability. Thus, tackling the indicators of knowledge economy at a national economy scale or at a partial scale, as in universities, requires identifying the components of knowledge economy and innovation in particular.

Innovation, previously mentioned, is a process of developing and improving the existing products or creating brand-new ones, in addition to enhancing processes and ways of production, services, business, or organizational models. In this sense, innovation is the basis for the knowledge-based economy and the main driver of the process of economic growth.

Innovative activities can be divided into four types: product-associated activities (technical in nature); activities associated with trademarks and distribution channels, with a product-delivery nature; activities associated with knowledge use or customer relationship management system, related to internal processes of organizational and institutional facilities; and activities associated with business models.

We can conclude that innovation has several forms including new products (goods and services), production processes, business practices, organizational and business models. And there are innovations directed at achieving social gains, known as "social innovations."

Economists are convinced day after another of the importance of innovations' role in the process of economic growth at the expense of capital accumulation in the long term. According to The Organization for Economic Co-operation and Development "OECD," the motive for increased economic development and high standard of living following the Second World War is represented in the rapid progress in technology and innovation. It is also estimated by USA commercial service that more than 75% of USA economic growth was driven by technological innovations since the Second World War until recently. A study on 98 developed and developing countries showed that around 90% of the real growth of average income per capita is attributable to innovation [12].

Fig. 1.1 Processes of knowledge at the university

Innovation leads to economic growth through providing institutions with new competitive advantages, raising productivity of traditional elements of production, opening broader areas for new investments and advanced products, increasing employment opportunities, rising wages by promoting productivity, and leading growth, employment, and wage rise toward prosperity.

In light of the local and global challenges faced by economy in the twenty-first century, countries have been calling for working hard on increasing competitiveness for their economy by focusing on knowledge as a main element in development. As for the Arab world, including Saudi Arabia, the Kingdom's Ninth Development Plan spotlighted the importance of increasing expenditures on scientific research up to the global levels to achieve development goals which are possible in case of having an innovation and knowledge-based economy where institutions, especially universities, should play a vital role in increasing their outcomes of knowledge, innovations, and technology development at the same global levels.

In order to perform their role, universities have to break out of the box of traditional ways and methods of teaching and scientific research which limit knowledge output and start focusing on creativity and innovation processes, utilizing modern techniques to improve the quality and quantity of the process of knowledge generation, production and spreading in a way that serves practical life and development progress for communities [13]. They are also required to provide a clear-objective mission at every level; pushing their members toward teamwork commitment; avoiding individualism; evaluating the effectiveness of educational process and university outputs; in addition to continuous interaction with business sectors for community development and enhancing mutual interest; and direct effective contribution in serving community, solving its problems, and fulfilling its recurring needs, as well as contributing to community development and knowledge increase.

1.7   Innovation in Saudi Arabia' Economy

Being important for world economies, innovations have also a special importance for Saudi economy that has been based on utilizing oil as a natural resource of income and wealth in community for many years since the discovery of oil until now. Hence, the Saudi economy's need for building a diverse economic base focused on high-technology activities has become an urgent priority for diversifying income and wealth resources, driving economic growth, and providing high-productivity and high-income job opportunities.

Hence, it is important to establish partnership between universities and business sectors in the Kingdom, as the

challenge today is the establishment of an economy based on modern technology and knowledge, then establishing projects in different economic fields, especially modern manufacturing, that depend on modern inputs, innovations, and technology, or knowledge-acquainted or highly skilled human resources. Innovation and modern technologies not only provide new products, but also establish new industries, which in turn increase added value, diversifying economy, creating new high-productivity and high-income job opportunities, and increasing competitiveness for local industries.

Technology support institutions have many advantages relating to spreading knowledge and promoting investment in national economy. At the beginning of the current decade (2011–2020), business incubators in North America provided job opportunities for eight thousand people, and an annual income of more than 7 billion US dollars, while in Europe, business incubators created about forty thousand job opportunities annually [14].

King Abdulaziz University, a pioneering university in KSA

Given the aforementioned, the government of the Custodian of the Two Holy Mosques has been seeking to achieve comprehensive development that would put the country on the list of developed countries by bringing about a broad knowledge leap and building three complimentary pillars: strong economy not based on complementary oil. Therefore, the Kingdom issued Vision 2030 [15]: A vibrant society, a thriving economy, and an ambitious nation. These pillars are coherent for achieving the objectives and maximizing the use of the vision's pillars. For community is where the vision starts and ends. The first pillar, emerging from faith in building up a community, represents the base of achieving this vision and establishing a solid foundation for the prosperity of the Kingdom's economy. In its second pillar (a thriving economy), the vision focuses on providing job opportunities for all by building an educational system related to labor market needs and developing opportunities for all, starting from entrepreneurs and small-scale enterprises to large companies. It also stresses developing the Kingdom's investment tools to unleash the potential of the promising economic sectors, diversify economy, generate job opportunities for citizens, and revitalize competitiveness leading to raising the quality of services and economic development, whereas efforts are focused on privatizing government services and enhancing business environment which contributes to attracting the best global competencies and specific investments in addition to utilizing the Kingdom's unique strategic location. The vision's third pillar concentrates on public sector, where it outlines the active government through enhancing efficiency, transparency, and accountability; encouraging culture of performance to empower human energy and resources; and preparing the environment required for citizens and business and non-profit sectors to take their responsibilities; in addition to taking the initiative in face of challenges and seizing opportunities.

Saudi Arabia's Vision 2030 targeted many objectives, including education and its consequent economic growth, such as:

1. Enabling five Saudi universities at least to be among the world's top (200) universities by 2030,
2. Providing educational opportunities for all in a suitable environment in light of the educational policy of the Kingdom,
3. Raising the quality of educational outputs,
4. Increasing the effectiveness of scientific research,
5. Encouraging creativity and innovation,
6. Developing community partnership.

1.8 Experiments of Pioneering Innovative Universities

Here, we briefly present some successful experiments of world's pioneering universities in terms of creativity and innovation. Developed countries around the world managed to maintain the momentum toward further development and economic progress through building innovative capacities and advanced technology, and forming communication and linking networks between innovation system stakeholders represented in universities, research centers, business sector, and government that helped in building an environment capable of transforming ideas into successful outputs, enabling them of producing new commodities and services and creative economic values. These experiments include:

1.8.1 Massachusetts State Universities in USA

USA state of Massachusetts has nine research universities; on top of them are Massachusetts Institute of Technology (MIT) and Harvard University [16], which are placed among top three universities in the world, in addition to other seven research universities listed among the world's top 100 universities [17–19]. In their local and national community, these universities push economic growth through their impact upon the state of Massachusetts because of their various contributions to developing the region in many technological fields.

This impact is represented in the following:

1. Obtaining patents granted by world offices,
2. Granting licenses annually for establishing new companies that would impact economic growth process,
3. Providing services and facilitation to educate senior graduates and workers and train them to acquire skills to do advanced jobs,
4. Providing offices and laboratories to support the expertise of new businesses, encouraging them to grow.

Historically, Massachusetts is considered a center for entrepreneurship and innovation. Its research universities contributed to establishing many large companies, which are considered important partners for the university, such as Baojun. It also participated, in cooperation with its partners, in establishing hundreds of companies in the areas surrounding the universities. In 2000 particularly, universities helped in establishing 41 companies for business investment of technologies produced at these universities. They also granted 280 licenses for special projects in 2000, where universities received $44.5 million. In addition, they also succeeded in attracting a number of global and national companies, to be settled in Boston (state capital), for developing basic and practical researches at the state.

Massachusetts Institute of Technology (MIT)

Boston University

research and innovation, providing required facilitation and services for innovation, helping foster and develop innovation and competitiveness in industry, and providing job opportunities for competent and distinguished individuals. In addition, they foster and help small and medium-sized enterprises (SMEs) in high-technology business fields. As a result, economy would be transformed from a natural-resource-based economy into innovation and technology-based one [21].

Innovation-based economy has become a powerful factor, pushing outstanding universities toward changing the policies of scientific research, as universities seek to direct their support to generate, encourage, and provide fund for creators and innovators. Thus, education and researches in universities are expected to be greatly impacted by knowledge-based industrial companies, whether established inside or outside universities [22]. In the USA and Japan, experiments proved that cooperation between teaching staff and industrialists plays an important role in the success of these companies, as researchers and academics participation in industrial researches and consultancies provided for industrial bodies are key factors for their success in the environment of knowledge-based economy.

1.8.2 US Harvard University

Harvard University provides its students with a broad platform for arts and science, as the university incurs huge costs for achieving such broad excellence, as its annual operational budget reaches $ 4 billion [20]. Chapter 6 of this book includes more information about Harvard University.

1.8.3 Science and Technology Parks

Scientific research and innovation system is an important cornerstone for knowledge societies, as it comprises a group of units mainly established by the state, with the participation of the private sector at some of which. These units are research centers, science parks, and business incubators. Research centers are located next to scientific departments in most universities around the world, but science parks, in spite of their establishment long history in the USA and other developed countries, have no specific definition, yet include the following components:

1. A real-estate area dedicated for the park.
2. An organizational structure for implementing activities of technology transfer.
3. Partnerships between higher educational institutions, government and private sector.

Innovative universities pay attention for establishing start-ups or technology incubators and providing required facilitation, which turns some of their students, graduates, and teaching staff into business pioneers, while these start-ups work on employing graduates and applying innovations. Thus, science and technology parks have become effective tools in development through focusing on scientific

1.8.4 Start-Ups

A start-up company is a newly established company in the first stage of its operations, development, and search for *markets*. Start-up founders effectively develop a scalable business model. Hence, the concepts of start-ups and entrepreneurship are similar. However, entrepreneurship refers to all new businesses, including *self-employment* and businesses that never intend to grow big, while start-ups refer to new businesses that intend to grow beyond the solo founder, have employees, and intend to grow large, but at the same time, only few companies go on to be successful, large, and influential. The simple launch of the start-up through self-funding or external funding with low initial capital makes it attractive for investors and founder entrepreneurs. In addition, a start-up is characterized by rapid growth, and in spite of the inherent high risk, it has high potential returns [23].

Start-ups depend on many funding sources, such as:

1. Self-funding,
2. Individual investors,
3. Venture capital companies.

Start-ups design a business model before their launch. This model is supposed to be unique because distinctiveness and excellence of ideas are the key factors of a start-up success and its most important means of achieving growth and expansion [24].

Ranking Methodology of the World's Most Innovative Universities

2.1 Ranking Organization of the World's Most Innovative Universities

Reuters News Agency started the idea of ranking the World's Most Innovative Universities. It relied on data compiled by Clarivate Analytics and several of its research platforms [25]. The latter company used to be called "Intellectual Property & Science Business of Thomson Reuters." The process began by identifying approximately 600 academic and government organizations that published the greatest number of articles in scholarly journals. Reuters, then, put ten criteria for the innovation to evaluate the chosen organizations. These criteria are presented later in Sect. 2.3.

2.2 Data Collection Methodology

Reuters relied on data compiled by Clarivate Analytics and several of its research platforms [25]. The platforms are:

1. InCites,
2. Web of Science,
3. Derwent Innovations Index,
4. Derwent World Patents Index,
5. Patents Citation Index.

Reuters began by identifying approximately 600 academic and government organizations that published the greatest number of articles in scholarly journals from 2011 to 2016, as indexed in the Clarivate Web of Science Core Collection database. The list was cross-referenced against the number of patents filed by each organization during the same time period in the Derwent World Patents Index and the Derwent Innovations Index. Patent equivalents, citing patents, and citing articles were included up to March 2018. The timeframe allows for the articles and patent activity to receive citations, thereby contributing to that portion of the methodology.

Next, the list was reduced to just those institutions that filed 70 or more world (World Intellectual Property Organization, WIPO) patents, the bulk of which were universities. The restriction to ranking only those institutions with 70 or more world patents is different for this global ranking than for our regional rankings.

For the 2018 surveys of Europe's Most Innovative Universities and Asia's Most Innovative Universities, the threshold was 50 world patents, because it allowed a more in-depth view of the most active institutions relative to others in their geographic area.

Each candidate university was then evaluated using various indicators including how often a university's patent applications were granted, how many patents were filed with global patent offices and local authorities, and how often the university's patents were cited by others. Universities were also evaluated in terms of how often their research papers were cited by patents and the percentage of articles that featured a co-author from industry.

Since some university systems, such as the University of California, centralize their patent administration, it is not possible to identify which of the various campuses was responsible for the initial research; so in these cases, the entire system was ranked, as opposed to an individual campus. Further complicating matters are the fact that not all universities publicly list their names on their patents or use complex name variants. For example, patents by the University of Oxford are filed under the name ISIS Innovations Ltd. In such instances, the name of the entity administrating patents for a university was identified and the patents were then associated with the university by the Clarivate analysts. There are several related changes to the global ranking methodology this year. In the past, Clarivate ranked the entire University of North Carolina together, instead of breaking out individual members, but in 2018 each member institution was ranked independently. Following this change, only one UNC institution placed in the top 100: The flagship University of North Carolina at Chapel Hill.

© The Author(s) 2021
The Leading World's Most Innovative Universities,
https://doi.org/10.1007/978-3-030-59694-1_2

2.3 Innovation Criteria

The following are the criteria that contributed to each university's composite score, which in turn determined the ranking of the universities according to innovative capacity and achievement [25]:

1. Patent Volume

Reuters used "Derwent World Patents Index" and "Derwent Innovations Index" to obtain the number of basic patents (patent families) filed by the organization. This is an indication of research output that has a potential for commercial value. The number is limited only to those patents that are registered with the World Intellectual Property Organization (WIPO).

2. Patent Success

Here also, Reuters used "Derwent World Patents Index" and "Derwent Innovations Index" to get the ratio of patent applications to grants over the assessed timeframe. This indicates the university's success in filing applications that are then accepted.

3. Global Patents

Here again, Reuters used "Derwent World Patents Index" and "Derwent Innovations Index" to obtain the percentage of patents for which coverage was sought with the US, European, and Japanese patent offices. Filing an international patent is an expensive and laborious process, and filing in multiple countries or regions is an indication that the invention is considered to be non-trivial and has commercial value.

4. Patent Citations

Reuters used "Patents Citation Index" to find the total number of times a patent has been cited by other patents. As part of the patent inspection process, the patent office examiner will cite significant prior art. The number of times a patent has been cited is an indication that it has an impact on other commercial R&D.

5. Patent Citation Impact

Here too, Reuters used "Patents Citation Index" to find the patent citation impact. This is an indication of how much impact a patent has had. Because it is a ratio (or average), it is not dependent on the size of the organization. Note that the indicator Percent of Patents Cited (listed below) is closely related to this indicator; therefore, these two indicators are given half the weighting of all others.

6. Percent of Patents Cited

Here again, Reuters used "Patents Citation Index" to find the percent of patent cited. This indicator is the proportion of patents that have been cited by other patents one or more times. As mentioned, it is closely tied to the Patent Citation Impact indicator.

7. Patent to Article Citation Impact

Reuters used the "Patents Citation Index," "Derwent World Patents Index," and "Web of Science Core Collection" to obtain the patent to article citation impact. This indicator, similar to Patent Citation Impact, measures the average number of times a journal article has been cited by patents. This unique indicator demonstrates that basic research conducted in an academic setting (as recorded in scholarly articles) has had influence and impact in the realm of commercial research & development (as measured by patents).

8. Industry Article Citation Impact

Here, Reuters used "Web of Science Core Collection" to get this impact. Article-to-article citations are an established indicator of influence and research impact. By limiting the citing articles only to those from industry, this indicator reveals the influence and impacts that basic research conducted in an academic setting has had on commercial research.

9. Percent of Industry Collaborative Articles

Here again, Reuters used "Web of Science Core Collection" to get the percentage of all articles of a university that contain one or more co-authors from a commercial entity. This indicator shows the percentage of research activity that is conducted in collaboration with industry, suggesting potential future economic impact of the research project jointly undertaken.

10. Total Web of Science Core Collection Papers

Here also, Reuters used "Web of Science Core Collection" to get the total number of journal articles published by the organization. This is a size-dependent measure of the research output of the university.

Table 2.1. Examples in the differences in ranking order between the global and Europe's lists (2018)

Institution	Rank in the global list	Rank in Europe's list
Imperial College London	8	2
Swiss Federal Polytechnic School of Lausanne	12	4
University of Cambridge	18	3
Pierre and Marie Curie University—Paris 6	–	18
University of Oxford	40	14
Technical University of Munich	45	6

Table 2.2. Examples in the differences in ranking order between the global and Asia-Pacific's lists (2018)

Institution	Rank in the global list	Rank in Asia-Pacific's list
Pohang University of Science And Technology (POSTECH)	13	3
University of Tokyo	20	2
Osaka University	22	6
Kyoto University	26	7
Seoul National University	34	4
Tohoku University	36	9

11. Final Score

The indicators were used to rank each of the universities, and the composite score was achieved by summing the ranks for each criterion for each university. Each indicator was weighted equally with the exception of Patent Citation Impact and Percent of Patents Cited, which received 50% weighting each since they are closely related in measuring the same phenomenon.

2.4 Tables of the Most Innovative Universities

Reuters has issued three different tables for the most innovative universities. These are:

1. The World's Most Innovative Universities, starting from 2015 and updated every year.
2. Europe's Most Innovative Universities, starting from 2016 and updated every year.
3. Asia-Pacific's Most Innovative Universities, starting from 2016 and updated every year.

It should be noted that an institution's ranking relative to others may be different in the global and the regional rankings since the ranking is dependent on summarizing the ranks of 10 indicators among others in the population, resulting in a composite score specifically geared to those in the comparison group. As mentioned above, in the global list, the institutions with 70 or more world patents were included, while, in Europe and Asia lists, the threshold was 50 world patents. Tables 2.1 and 2.2 show some examples of ranking orders in the global and the regional rankings.

3.1 Introduction to Ranking

Stanford University has been ranked as the World's Most Innovative University for the fourth year in a row according to the latest Reuters ranking (2018) [26].

Examining the top 10 universities in 2018 ranking, a slight change can be seen from 2017 ranking. That is to say, eight of 2017 top 10 world universities remained in the top of the 2018 list. Meanwhile, Imperial College London and The University of North Carolina at Chapel Hill joined 2018 top 10 list for the first time. In contrast, University of Michigan and Korea Advanced Institute of Science and Technology (KAIST) have fallen out of the list [26].

Also, Massachusetts Institute of Technology (MIT) and Harvard University maintained their ranks of 2017; as they were ranked the second and third, respectively US universities occupied about half of 2018 World University Ranking list, where 46 US top innovative universities were listed with three less universities than 2017.

KU Leuven University in Belgium was ranked seventh, (two places back compared to the fifth rank achieved in 2017) but still remains on top of the innovative European universities in 2018. This is partially because the large number of highly important patents the university introduces, making it the best university outside the USA. Imperial College London occupied the eighth place in 2018 ranking, up from the 15th in 2017 list, making it the first in UK and the second in Europe. Meanwhile, among the world top 100 universities, European universities occupied 27 positions (26 of which were already in 2017 list), mostly in UK and Germany.

Apart from the USA and Europe, Asia occupied 23 positions mostly in South Korea and Japan. Korea Advanced Institute of Science and Technology (KAIST) ranked the 11th in the world (back from the sixth in 2017). However, it is still the Most Innovative Asian University preceding the Japanese universities in this regard.

New economic tigers and the fierce competition shown by Japan, South Korea, and China are to be taken into consideration. However, the ranking of the World's Most Innovative Universities indicates that effective scientific research, which contributes to technological inventions and supports new industries and promising markets, is still conducted in the West not the East. The reason behind this is the fact that USA still largely funds scientific research, making Stanford University on top of the World's Most Innovative Universities. This university, located in the heart of California's Silicon Valley, continues to play a major role in the development of the modern world of Internet connections. Graduates of Stanford have established the world's most renowned technology companies such as Google and Intel. The university continues to conduct new research and develop up-to-date technologies, which makes its research an important source of knowledge for researchers all over the world, noting the high percentage of citation recorded.

According to the ranking of the World's Most Innovative universities, the top universities in this field are the most renowned and experienced universities of North America and Western Europe. Overall, American and Western European universities occupied 75 of the 100 places (48 for North American and 27 for Western European). The remaining 25 places are occupied by Asian universities, one of which is Japanese and two are South Korean, which introduce their programs in English. Among the top Asian universities is Korea Advanced Institute of Science and Technology (KAIST), established in 1971 by the Korean government on the model of US engineering faculties, and funded with a loan worth millions of dollars by the US Agency for International Development. On the other hand, none of the universities in Africa, South America, Oceania, or Arab countries, have been ranked in the innovative universities list yet.

It should be noted that the ranking of universities according to the ten criteria of innovation does not reflect the innovative efforts exerted by the researchers of a given department, faculty, or research center in a university. The ranking is based on the measurement of innovation at the university level as a whole, which means that an innovative

© The Author(s) 2021
The Leading World's Most Innovative Universities,
https://doi.org/10.1007/978-3-030-59694-1_3

department may exist in a particular university, but the university as a whole may not be included or listed at the end of the ranking list. This, of course, does not underestimate the efforts of such an innovative research center, which could serve as an effective source of modern science and technology.

3.2 Table of the World's Most Innovative Universities

Table 3.1 shows the World's Most Innovative Universities in 2018 [26]. This is the most recent rankings when this book was composed. In order to unify with the rankings of

Asia-Pacific's ranking table, Table 3.1 stopped at 75 most innovative universities.

From Table 3.1, it is clear that the highest number of the World's Most Innovative Universities in 2018 is in the USA. In Table 3.2, the numbers of the World's Most Innovative Universities in 2017 and 2018 in the different countries are given for comparison.

Table 3.1. World's most innovative universities in 2018

Rank	Institution	Country
1	Stanford University	USA
2	Massachusetts Institute of Technology (MIT)	USA
3	Harvard University	USA
4	University of Pennsylvania	USA
5	University of Washington	USA
6	University of Texas System	USA
7	KU Leuven	Belgium
8	Imperial College London	UK
9	University of North Carolina Chapel Hill	USA
10	Vanderbilt University	USA
11	Korea Advanced Institute of Science & Technology (KAIST)	South Korea
12	Ecole Polytechnique Federale de Lausanne	Switzerland
13	Pohang University of Science & Technology (POSTECH)	South Korea
14	University of California System	USA
15	University of Southern California	USA
16	Cornell University	USA
17	Duke University	USA
18	University of Cambridge	UK
19	Johns Hopkins University	USA
20	University of Tokyo	Japan
21	California Institute of Technology	USA
22	Osaka University	Japan
23	University of Michigan System	USA
24	Northwestern University	USA
25	University of Wisconsin System	USA
26	Kyoto University	Japan
27	University of Minnesota System	USA
28	University of Illinois System	USA
29	Georgia Institute of Technology	USA
30	University of Utah	USA

(continued)

Table 3.1. (continued)

Rank	Institution	Country
31	University of Erlangen Nuremberg	Germany
32	Ohio State University	USA
33	Columbia University	USA
34	Seoul National University	South Korea
35	University of Toronto	Canada
36	Tohoku University	Japan
37	University of Pittsburgh	USA
38	Yale University	USA
39	Sungkyunkwan University	South Korea
40	University of Oxford	UK
41	University of Colorado System	USA
42	Tufts University	USA
43	Baylor College of Medicine	USA
44	Tsinghua University	China
45	Technical University of Munich	Germany
46	Kyushu University	Japan
47	Tokyo Institute of Technology	Japan
48	University College London	UK
49	ETH Zurich	Switzerland
50	Purdue University System	USA
51	University of Chicago	USA
52	Oregon Health & Science University	USA
53	University of Manchester	UK
54	Indiana University System	USA
55	Universite de Montpellier	France
56	University of Munich	Germany
57	Technical University of Denmark	Denmark
58	Emory University	USA
59	Peking University	China
60	Sorbonne University	France
61	University of British Columbia	Canada
62	Delft University of Technology	Netherlands
63	National University of Singapore	Singapore
64	Princeton University	USA
65	University of Zurich	Switzerland
66	Hanyang University	South Korea
67	Case Western Reserve University	USA
68	Yonsei University	South Korea
69	Rutgers State University New Brunswick	USA
70	Boston University	USA
71	University of Massachusetts System	USA
72	Johannes Gutenberg University of Mainz	Germany
73	Wake Forest University	USA
74	Keio University	Japan
75	Korea University	South Korea

Table 3.2. Numbers of the world's most innovative universities in the different countries

Countries	Numbers of the world's most innovative universities	
	2017	2018
USA	40	39
Japan	7	7
South Korea	6	7
UK	4	5
Germany	3	4
Switzerland	3	3
France	2	2
Canada	2	2
China	2	2
Belgium	2	1
Denmark	2	1
The Netherland	1	1
Singapore	1	1
Total	**75**	75

4.1 Introduction to Ranking

In ranking of 2018 European Most Innovative Universities for the year 2018, Belgian University of KU Leuven topped the list, becoming the first innovative university in Europe for the third year in a row, among the universities that work on the advancement of science, development of modern technologies, and support of industries and markets. It should be noted that this University, established in 1425, offers its programs in Dutch. Then comes Imperial College London in the second place and Cambridge University in the third, maintaining their ranking for the third year in a row. However, the ranks of some other universities have changed in 2017 and 2018 lists.

Although the number of listed German universities (23 universities) has not changed in 2017 and 2018, many of them have achieved better ranks in the 2018 list. Cumulatively, it can be said that they are collectively 23 ranks up, which is the greatest progress achieved by a European country in the list. British universities have increased from 17 in 2017 to 21 in 2018, but cumulatively they have dropped 35 places against universities from other countries. Perhaps, the most obvious reason behind this is the withdrawal of the UK from the European Union, briefly expressed as "**Brexit**". Although it is still under discussion and will take time to be finalized, it may lead some scientists to move from Britain to other European countries. A study undertaken in February 2018 by Center for Global Higher Education, headquartered in UK, suggests that many German universities consider the Brexit as an advantage, for it enables German universities to attract British researchers. Meanwhile, the study suggests that academic staff in the UK complain about the fewer number of British postdocs who have already started to look for jobs in the European Union or the USA, something that is expected to get worse upon **Brexit**. A study undertaken in November 2017 by School of International Futures for the Royal Society in the UK predicts that British universities will compete for fewer qualified workers and projects that used to receive European funds. British researchers invited to contribute to joint research projects and conferences will also decrease as well as the number of international cooperation agreements. Similarly, European institutions funding research in universities may prefer to keep investments in the European Union to avoid taxes and problems that may arise from working with British universities after **Brexit** [28].

On the other hand, Germany has significantly focused on supporting sciences, and increased research budgets on federal government level in order to support the growth of new industries, such as renewable energy. This priority aligns with the fact that German Chancellor Angela Merkel holds a doctorate in quantum chemistry and has worked as a scientific researcher before getting involved in politics. Based on 2017 analysis published in Nature Magazine, researchers are attracted to come and work in Germany because it allocates a budget of EUR 4.6 billion for the Excellence Initiative, which has attracted almost 4000 foreign scientists to Germany since 2005. In 2016, German research organization, Deutsche Forschungsgemeinschaft (DFG), the main funding agency in Germany, allocated 2.9 EUR billion for research scholarships, which means the increasing probability of supporting research proposals with a considerable rate.

2018 list of top innovative European universities indicates that there is no relationship between the number of distinct universities in a given country and its size. For instance, Belgium has 7 universities listed, while it has a population of only 11 million. The same applies to Switzerland, Denmark, the Netherlands, and the Republic of Ireland. While Russia, the most populous country on the European continent and the fifth largest economy in Europe, does not have any universities in this list.

Either a university is ranked on the top or in the bottom of this list; the university remains among the top 100 universities in Europe. These universities are the source of original research and useful technologies that motivate and support the world economy.

The Leading World's Most Innovative Universities,
https://doi.org/10.1007/978-3-030-59694-1_4

4.2 Table of the Most Innovative Universities in Europe

Table 4.1 shows the Europe's Most Innovative Universities in 2018 [28]. This is the most recent rankings when this book was composed. In order to unify with the rankings of Asia–Pacific's ranking table, Table 4.1 stopped at 75 most innovative universities.

From Table 4.1, it is clear that the highest number of the Europe's Most Innovative Universities in 2018 is in Germany. In Table 4.2, the numbers of Europe's Most Innovative Universities in 2017 and 2018 in the different countries are given for comparison.

Table 4.1 Europe's most innovative universities in 2018

Rank	Institution	Country
1	KU Leuven	Belgium
2	Imperial College London	UK
3	University of Cambridge	UK
4	Federal Institute of Technology in Lausanne (EPFL)	Switzerland
5	University of Erlangen Nuremberg	Germany
6	Technical University of Munich	Germany
7	University of Manchester	UK
8	University of Munich	Germany
9	Technical University of Denmark	Denmark
10	Swiss Federal Institute of Technology Zurich	Switzerland
11	University College London	UK
12	Delft University of Technology	Netherlands
13	University of Zurich	Switzerland
14	University of Oxford	UK
15	University of Basel	Switzerland
16	University of Montpellier	France
17	Leiden University	Netherlands
18	Pierre & Marie Curie University—Paris 6	France
19	University of Paris Descartes—Paris 5	France
20	Ruprecht Karl University Heidelberg	Germany
21	Johannes Gutenberg University of Mainz	Germany
22	Free University of Berlin	Germany
23	Eindhoven University of Technology	Netherlands
24	University of Freiburg	Germany
25	University of Paris Sud—Paris 11	France
26	Charité Medical University of Berlin	Germany
27	Humboldt University of Berlin	Germany
28	Grenoble Alpes University	France
29	Dresden University of Technology	Germany
30	University of Bordeaux	France
31	Karlsruhe Institute of Technology	Germany
32	University of Oslo	Norway
33	Ghent University	Belgium
34	University of Birmingham	UK
35	University of Claude Bernard—Lyon 1	France
36	University of Glasgow	UK

(continued)

Table 4.1 (continued)

Rank	Institution	Country
37	Queen Mary University London	UK
38	King's College London	UK
39	Technical University of Berlin	Germany
40	RWTH Aachen University	Germany
41	University of Strasbourg	France
42	Free University of Brussels	Belgium
43	University of Copenhagen	Denmark
44	Polytechnic University of Milan	Italy
45	University of Edinburgh	UK
46	Grenoble Institute of Technology	France
47	Vrije University of Brussels	Belgium
48	Utrecht University	Netherlands
49	Ecole Polytechnique	France
50	Goethe University Frankfurt	Germany
51	University of Paris Diderot—Paris 7	France
52	University of Munster	Germany
53	Cardiff University	UK
54	Catholic University of Louvain	Belgium
55	Hannover Medical School	Germany
56	Erasmus University Rotterdam	Netherlands
57	University of Amsterdam	Netherlands
58	University of Dundee	UK
59	University of Aix-Marseille	France
60	University of Leicester	UK
61	Saarland University	Germany
62	University of Sheffield	UK
63	Eberhard Karls University of Tubingen	Germany
64	Vienna University of Technology	Austria
65	Trinity College Dublin	Ireland
66	University of Milan	Italy
67	University of Paul Sabatier—Toulouse III	France
68	University of Leeds	UK
69	University of Barcelona	Spain
70	University of Southampton	UK
71	University of Stuttgart	Germany
72	University of Wurzburg	Germany
73	University of Lorraine	France
74	University of Geneva	Switzerland
75	University of Twente	Netherlands

Table 4.2 Numbers of the Europe's most innovative universities in the different countries

Countries	Numbers of the world's most innovative universities	
	2017	2018
Germany	18	20
UK	15	16
France	15	14
The Netherland	7	7
Belgium	5	5
Switzerland	4	5
Denmark	3	2
Italy	3	2
Spain	1	1
Ireland	2	1
Norway	1	1
Austria	1	1
Total	**75**	75

5.1 Introduction to Ranking

Despite of the political instability in South Korea, there are strong and solid relations between universities and industry. These relations continue to lead economic growth and technical innovation in this country. This is the conclusion reached by Reuters in the third annual classification of Asian and Pacific universities, working on achieving progress in sciences and creating new technologies [29]. Korea Advanced Institute of Science and Technology, currently known as KAIST, is ranked the first for the third year in a row. Historically speaking, KAIST is the oldest Korean university dedicated for research, sciences, and engineering. It has three branch campuses in the following cities: Daejeon, Seoul, and Busan. The university produces a large number of innovations and applies for more patents than the other 75 universities on the list. In addition, researchers all over the world cite highly the research and patents of this university.

The second ranked on the Asia-Pacific's Most Innovative Universities list was occupied by the ancient Japanese University of Tokyo, which moved from the third (in 2017) to the second position (in 2018). Pohang University of Science & Technology (POSTECH) came in the third position, achieving a progress from the fourth position in 2017. It is worth mentioning that POSTECH was established by POSCO, a South Korean steel-making company, in 1986. POSTECH is the top university in terms of the number of scientific papers, submitted by industrial researchers, and the number of citations taken from POSTECH papers and found in scientific papers submitted by the private sector.

As for the fourth position, it was occupied by Seoul National University (SNU), which was established in 1946 as the first national university. It received a support worth USD 68 million (KRW76 billion) accounting for 15% of the total funds from external industries. It is known that universities in South Korea have close relations with industry. Classified the second in 2017, this university is two positions

back in the list. Meanwhile, Chinese Tsinghua University moved from the sixth (in 2017) to the fifth (in 2018).

In addition, three non-ranked Chinese universities have joined the 2018 list of Asia-Pacific that includes 75 universities increasing the number of innovative Chinese universities to 27, three of which are located in Hong Kong. Meanwhile, the number of South Korean listed universities dropped from 22 (in 2017) to 20 (in 2018). However, the number remains more than expected from a country with a population of less than 51 million people, compared to China whose population is more than 1370 million people. On the other hand, Japan maintained its 19 universities in the ranking list for the second year in a row.

Although India's population has grown more than 1280 million people, only one Indian university, the Indian Institutes of Technology (IIT), joined the list. The institute is a network of 23 universities with a central management of patents, which makes it difficult to know to which university a research paper belongs.

Countries with no ranking in the list include: Indonesia, Pakistan, and Bangladesh, the third, fourth, and fifth Asian countries in terms of population, respectively. In addition, there are no universities from either the Arabian Gulf, the Philippines, or Vietnam in the list, despite of their huge economies [29].

5.2 Table of Asia-Pacific's Most Innovative Universities

Table 5.1 shows the Asia-Pacific's Most Innovative Universities in 2018 [29]. This is the most recent rankings when this book was composed. It lists the 75 Most Innovative Universities in this region.

From Table 5.1, it is clear that the highest number of the Asia-Pacific's Most Innovative Universities in 2018 is in China. In Table 5.2, the numbers of Asia-Pacific's Most Innovative Universities in 2017 and 2018 in the different countries are given for comparison.

© The Author(s) 2021
The Leading World's Most Innovative Universities,
https://doi.org/10.1007/978-3-030-59694-1_5

Table 5.1. Asia-Pacific's most innovative universities in 2018

Rank	Institution	Country
1	Korea Advanced Institute of Science and Technology (KAIST)	South Korea
2	University of Tokyo	Japan
3	Pohang University of Science and Technology (POSTECH)	South Korea
4	Seoul National University	South Korea
5	Tsinghua University	China
6	Osaka University	Japan
7	Kyoto University	Japan
8	Sungkyunkwan University	South Korea
9	Tohoku University	Japan
10	National University of Singapore	Singapore
11	Hanyang University	South Korea
12	Peking University	China
13	Yonsei University	South Korea
14	Kyushu University	Japan
15	Korea University	South Korea
16	Tokyo Institute of Technology	Japan
17	Fudan University	China
18	Keio University	Japan
19	Shanghai Jiao Tong University	China
20	Gwangju Institute of Science and Technology	South Korea
21	Zhejiang University	China
22	Chinese University of Hong Kong	Hong Kong
23	Hokkaido University	Japan
24	Kyung Hee University	South Korea
25	Monash University	Australia
26	Nanyang Technological University	Singapore
27	Ajou University	South Korea
28	Huazhong University of Science and Technology	China
29	Hiroshima University	Japan
30	Kumamoto University	Japan
31	Nagoya University	Japan
32	Beijing University of Chemical Technology	China
33	East China University of Science and Technology	China
34	Tokyo Medical & Dental University (TMDU)	Japan
35	Tianjin University	China
36	University of Sydney	Australia
37	Ewha Womans University	South Korea
38	Hong Kong University of Science and Technology	Hong Kong
39	University of Auckland	New Zealand
40	Shinshu University	Japan
41	South China University of Technology	China
42	University of Queensland	Australia
43	Kanazawa University	Japan
44	China University of Petroleum	China
45	University of Melbourne	Australia
46	Southeast University China	China

(continued)

Table 5.1. (continued)

Rank	Institution	Country
47	University of Hong Kong	Hong Kong
48	University of Tsukuba	Japan
49	Catholic University of Korea	South Korea
50	Nanjing University	China
51	University of New South Wales Sydney	Australia
52	Chiba University	Japan
53	Xi'an Jiaotong University	China
54	Chonbuk National University	South Korea
55	Chonnam National University	South Korea
56	China University of Mining And Technology	China
57	Pusan National University	South Korea
58	Dalian University of Technology	China
59	Kayama University	Japan
60	Kyungpook National University	South Korea
61	Harbin Institute of Technology	China
62	Nankai University	China
63	Chung-Ang University	South Korea
64	Inha University	South Korea
65	Sun Yat-sen University	China
66	Sichuan University	China
67	Shandong University	China
68	University of Electronic Science and Technology of China	China
69	University of Ulsan	South Korea
70	Waseda University	Japan
71	Indian Institutes of Technology System (IIT)	India
72	Kobe University	Japan
73	Konkuk University	South Korea
74	Xiamen University	China
75	Tongji University	China

Table 5.2. Numbers of the Asia-Pacific's most innovative universities in the different countries

Countries	Numbers of the world's most innovative universities	
	2017	2018
China (Including Hong Kong)	24	27
South Korea	23	20
Japan	19	19
Australia	5	5
Singapore	2	2
New Zealand	1	1
India	1	1
Total	**75**	75

A Study of Some of the World's Most Innovative Universities

This chapter examines in detail a study of the top twenty most innovative universities in the world according to Table 3.1. A study of one university of each country, whose name was not included among the top twenty universities, is also added. The first-ranked university in each country which none of its universities were ranked among the top twenty is selected, and its details described the same way as that of the top twenty universities. The ranking of the added universities is: 31, 35, 44, 55, 57, 62, and 63. This study is shown below.

Stanford University is located in the heart of California's Silicon Valley. It was founded in 1891 by California senator and railway magnate Leland Stanford as a memorial of his deceased son, with the promise that "the children of California shall be our children", said the senator. Stanford University, a private research university takes first place on Reuters' list of the World's Most Innovative Universities for the third consecutive year, holding fast to its ranking by consistently producing new patents and papers that influence researchers around the globe [30].

6.1 Stanford University—USA

The Leading Worldâ s Most Innovative Universities,
https://doi.org/10.1007/978-3-030-59694-1_6

In July 2017, the university opened the Stanford Center for Definitive and Curative Medicine, a new research center focused on the development of stem cell and gene therapies to treat millions of people with genetic diseases worldwide. Also, the university recently opened the Stanford Laboratory for Cell and Gene Medicine, a $35 million, 23,000-ft^2 manufacturing facility for making stem-cell-based therapies for use in human patients.

In the 2016/2017 fiscal year, Stanford announced a total research budget of $1.6 billion. There are more than 6000 externally sponsored projects throughout the university. The federal government sponsors approximately 81% of these projects, including National Accelerator Laboratory (SLAC), while the rest of the research budget comes from other funding sources.

SLAC National Accelerator Laboratory is actually a U.S. Department of Energy laboratory, but is operated by Stanford University. It has a variety of programs, including programs in the atomic, physics, biology, and medicine fields.

Recent researches in the university tackle the development of a drug delivery device that monitors drug levels in the body real time and delivers proper dosages to the patient. Stanford researchers have also created a vine-like, flexible "growing" robot that operates in the narrowest places, and a new camera that allows robots to capture 4D images in a 140° field of vision. This 4D image allows viewers to refocus the image after it is captured, allowing processing the image after being taken.

Stanford University has played a key role in the development of the network-connected world. In the early 1970s, Stanford professor Vint Cerf co-designed the TCP/IP protocol that became the basic standard means of communication in the Internet. In 1991, physicists at the Stanford Linear Accelerator Center deployed the first World Wide Web server outside of Europe.

The Stanford faculty and alumni have founded major tech companies including Google, Hewlett-Packard (HP), and Cisco Systems. A 2012 study by the university estimated that companies formed by Stanford entrepreneurs generate so much revenue that if they were of an independent nation, it would rank among the ten largest economies in the world. Table 6.1 displays some key statistical figures of Stanford University that illustrate the quality of research and innovation [30–32].

Figure 6.1 shows the patents submitted by Stanford University in various fields, while Fig. 6.2 shows the percentage of accepted patents. Figure 6.3 shows the university's ranking per each of the ten innovation indicators [33].

Table 6.1 Some key statistical figures of Stanford University

Number of Nobel laureates among professors and graduates	81
Number of professors or graduates who are Turing Prize winners	27
Number of professors or graduates who are Fields Medal winners	7
Endowments in 2017	$22.8 billion
Campus area	33.1 km^2
Number of applicants for admission to the university (2016)	43,997
Number of students admitted to the university (2016)	2118 (5% of applicants)
Number of undergraduate students (2016)	7032 (43% of total students)
Number of postgraduate students (2016)	9304 (57% of total students)
Teaching staff	2153
Members of the academic council	4082
Student-to-academic staff ratio	1:4
Number of administrative staff	12,148

Fig. 6.1 Patents submitted by Stanford University

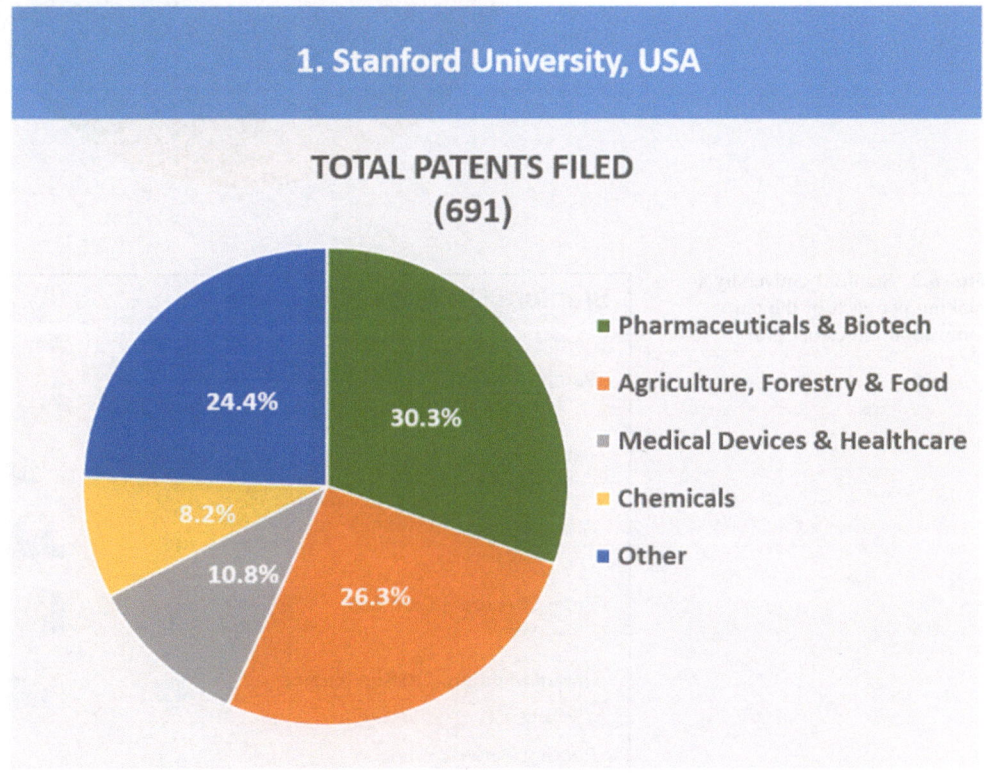

Fig. 6.2 Percentage of accepted patents of Stanford University

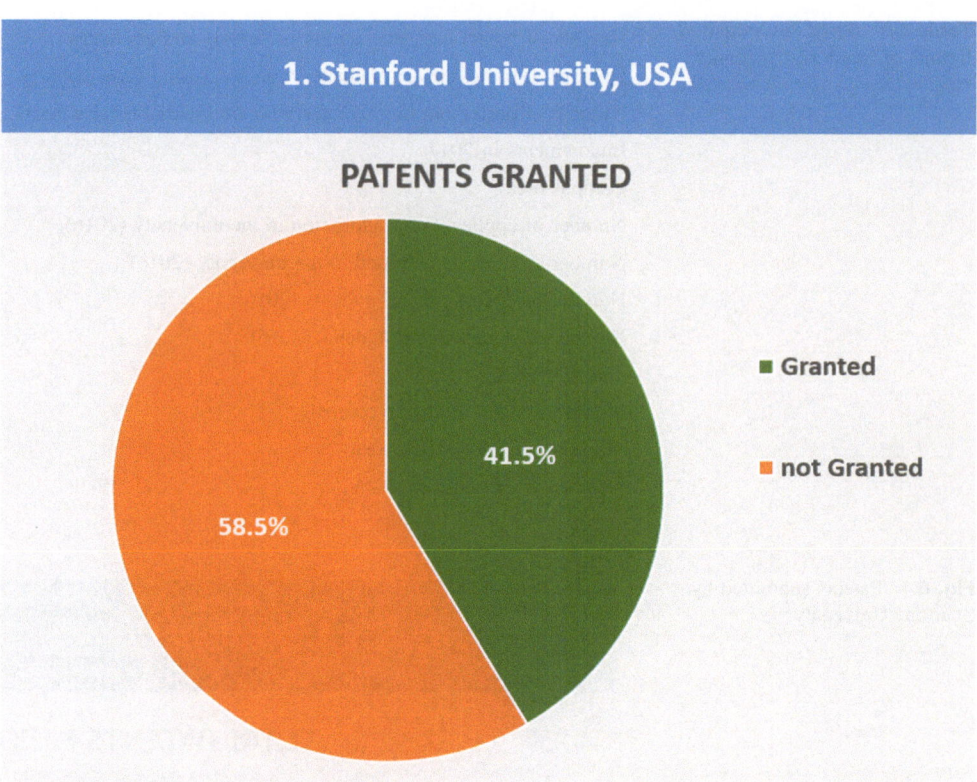

Fig. 6.3 Stanford University's ranking per each of the ten innovation indicators [33]

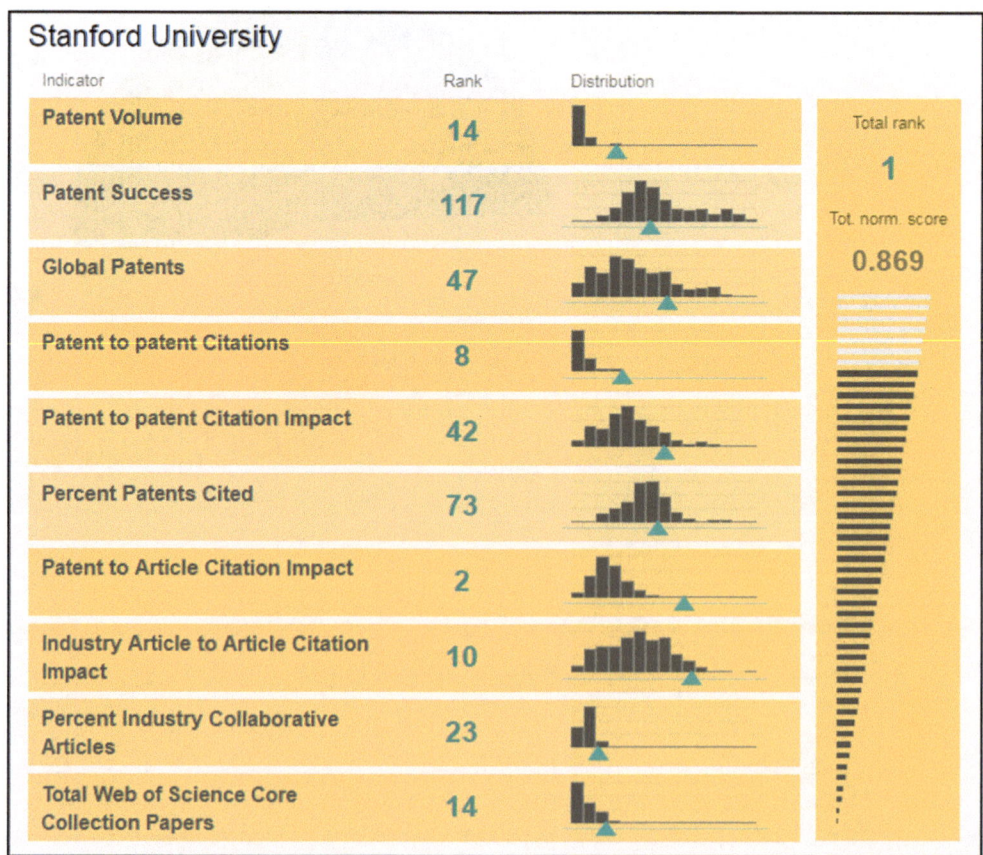

6.2 Massachusetts Institute of Technology (MIT)—USA

Massachusetts Institute of Technology (MIT), a private technological research university, is the second most innovative university in the world for the third consecutive year [34]. It was founded in 1861 with the mission of "enhancing the advancement, development and practical application of science." MIT research has fueled some of the most important innovations of the past century, including the development of digital computers and the completion of the Human Genome Project. Throughout its 155-year history, 89 of its professors and alumni were Nobel Prize laureates. Corporations founded by MIT's staff and alumni include: Bose, Dropbox, and iRobot. A 2014 study conducted by the university revealed that about 30,000 of these companies employed 4.6 million people and produced annual revenue of approximately $1.9 trillion.

In the fiscal year 2016, MIT research expenditures exceeded $728 million, 19% of which came from industry sponsors, including Ford, Intel, Shell, and Samsung. Approximately 700 companies work with MIT faculty and

Table 6.2 Some key statistical figures of Massachusetts Institute of Technology

Number of Nobel laureates among professors and alumni	89
Number of professors or graduates who are Turing Prize winners	15
Number of Fields Medalists among professors or alumni	6
Number of National Medal of Science winners	58
Number of National Medal of Technology and Innovation winners	29
Number of MIT alumni astronauts	34
Endowments in 2017	$14.8 billion
Campus area	68 ha
Number of applicants for admission to the university (2016)	20,247
Number of students admitted to the university (2016)	1452 (7.2% of applicants)
Number of undergraduate students (2017)	4547 (39.7% of total students)
Number of Postgraduate Students (2017)	6919 (60.3% of total students)
Number of international postgraduate students (2017)	2868 (41.5% of total students)
Teaching staff	1047
Members of the academic council	5893
Student-to-academic staff ratio	1:4
Total number of university staff	12,607

Fig. 6.4 Patents submitted by Massachusetts Institute of Technology

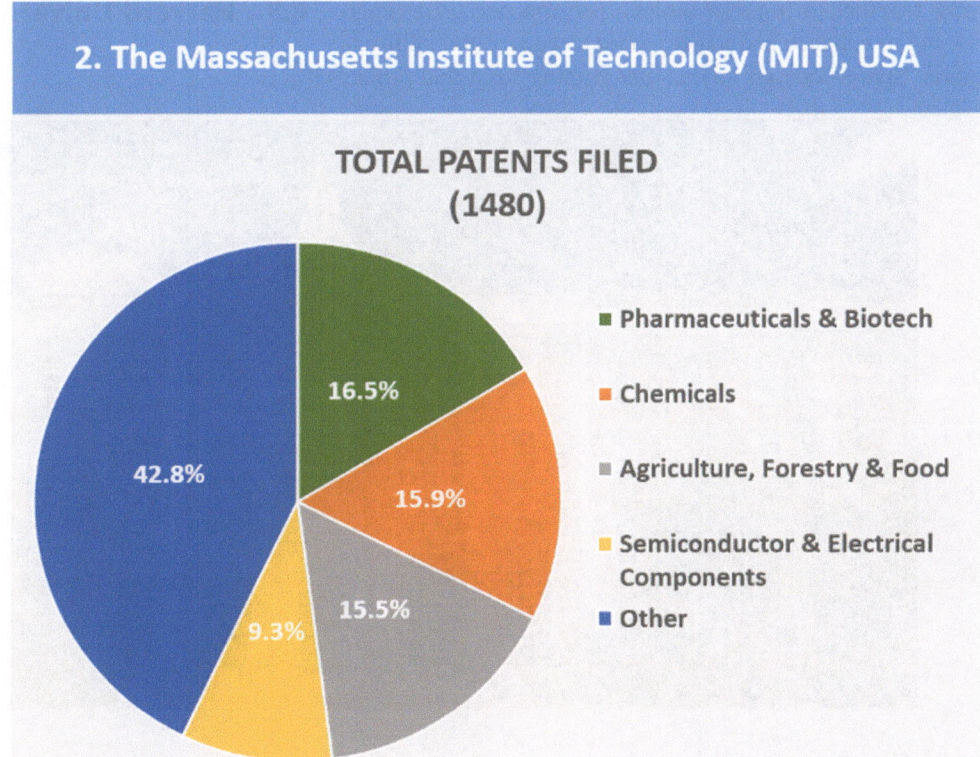

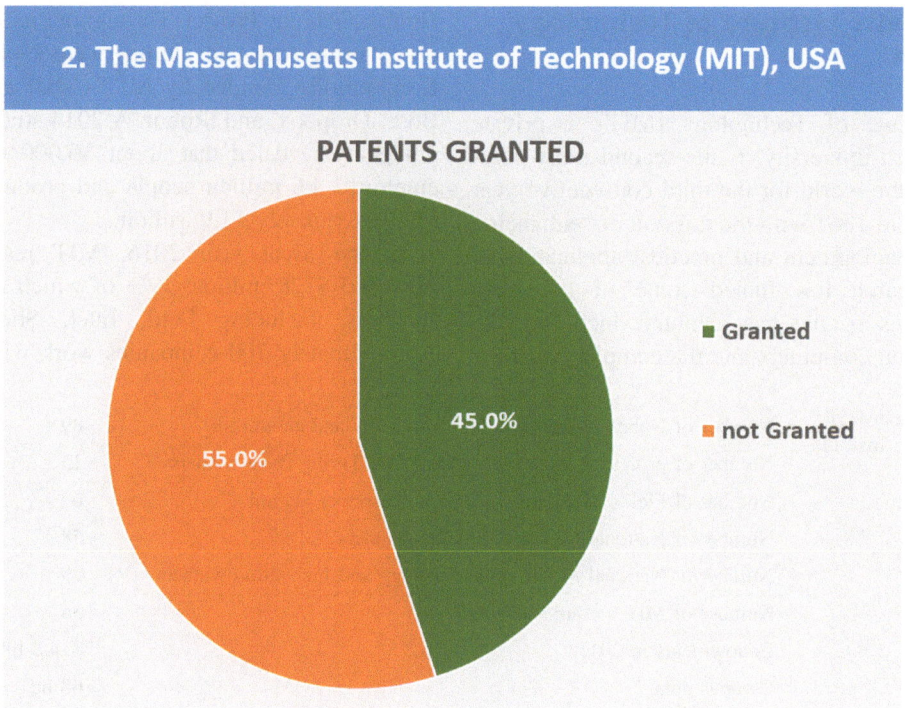

Fig. 6.5 Percentage of accepted patents of Massachusetts Institute of Technology

students through various programs such as the MIT Energy Initiative and the Industrial Liaison Program.

In 2011, the university launched an open-source platform that allows people across the globe to enroll in free online classes taught by MIT professors covering a wide variety of topics [35–36].

In 2017, MIT revamped its Sports Technology Group, which became MIT 3-Sigma Sports, a program connecting students and faculty with alumni and industry partners who work together to improve athletic performance using engineering to enhance endurance, speed, accuracy and agility in sports. Sports-focused innovations coming from this project are now in use by divers groups, International Cricket Council and CBS TV network, which uses the Emmy-award winning Swing Vision system in golf footage.

Table 6.2 displays some key statistical figures that illustrate the quality of research and innovation [34, 37–39]

Figure 6.4 shows the patents submitted by Massachusetts Institute of Technology in various fields, while Fig. 6.5 shows the percentage of accepted patents.

6.3   Harvard University—USA

Harvard University, a private university, is ranked third among the World's Most Innovative Universities for the third consecutive year [40]. Established in 1636, Harvard University is the oldest institution of higher education in the USA. It has presented 157 Nobel laureates, 32 heads of state, and 48 Pulitzer Prize winners. With more than $37 billion, Harvard has the largest endowment in the world. More than one million students from 195 countries around the world have enrolled in the university's free online programs [41].

In the 2014/2015 fiscal year, Harvard had approximately $800 million budget for sponsored research. Nearly $50 million of which came from corporate sponsors. That budget currently exceeds $900 million.

In 2016, Harvard researchers developing quantum computing technology built a radio receiver the size of two atoms, and in 2017, researchers at the university's Wyss Institute for Biologically Inspired Engineering created a biological hard drive using gene-editing technology to alter cells in order to allow them to reserve and recall bits of DNA-encoded information. This approach could lead to new data storage formats, or could be engineered into memory devices able to record a cell's molecular activity during development or under exposure to pathogens. Researchers at the Wyss Institute also developed a tethered exosuit that can increase a runner's performance by reducing the metabolic cost of running. There are medical applications of this technology for the treatment of stroke victims and other patients who have reduced mobility.

Table 6.3 Some key statistical figures for Harvard University

Number of Nobel laureates among professors and alumni	157
Number of Turing Award winners among professors and alumni	14
Number of Fields Medalists among professors and alumni	18
Number of Pulitzer Prize winners among professors and alumni	48
Endowments in 2017	$37 billion
Campus area	85 ha
Number of applicants for admission to the university (2021)	39,506
Number of students admitted to the university (2021)	2037 (5.2% of applicants)
Number of undergraduate students	6700 (31.6% of total students)
Number of postgraduate students	14,500 (68.4% of total students)
Teaching staff	2400
Members of the academic council	4388
Total number of university staff	16,000

Fig. 6.6 Patents submitted by Harvard University

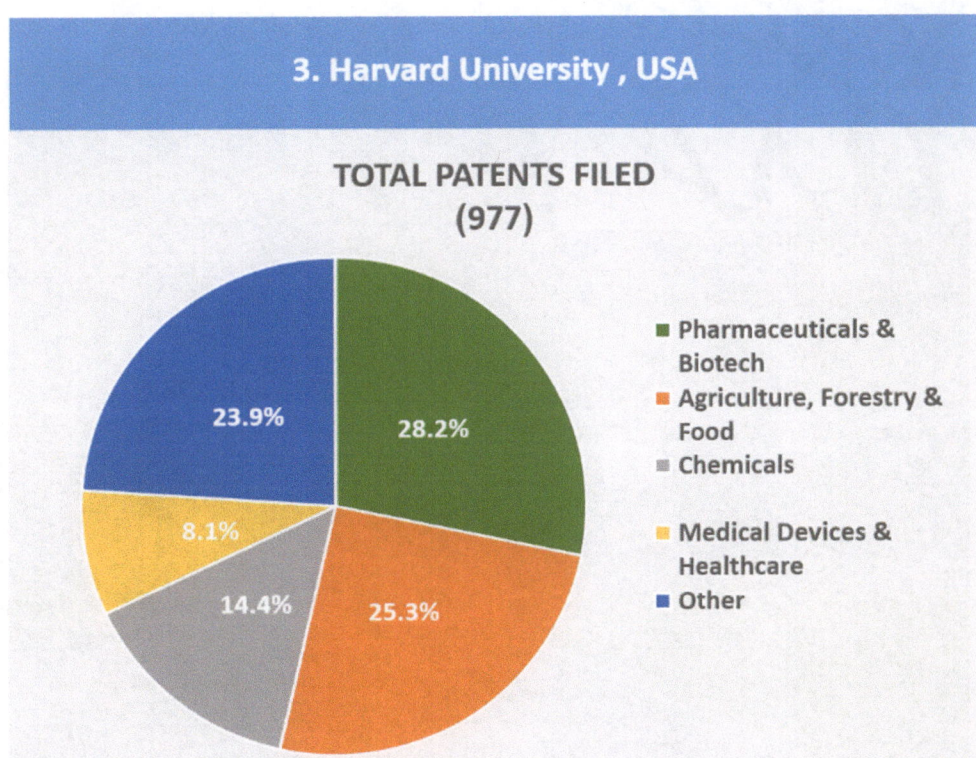

Fig. 6.7 Percentage of accepted patents for Harvard University

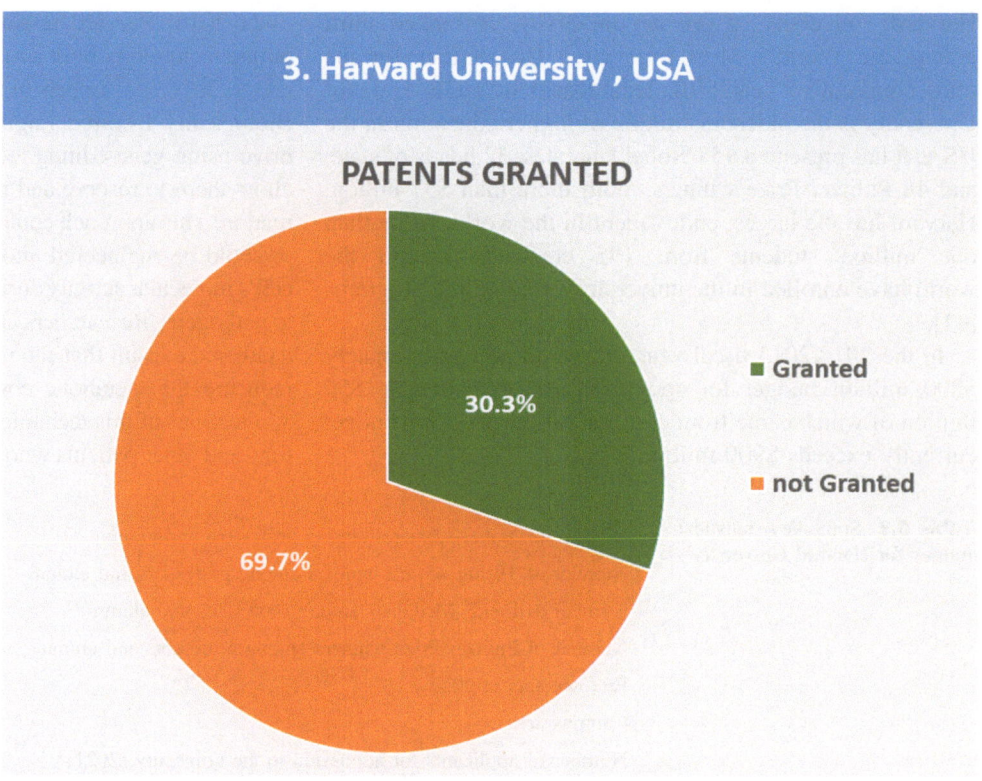

Table 6.3 displays some key statistical figures that illustrate the quality of research and innovation [40, 42–44].

Figure 6.6 displays the patents submitted by Harvard University in various fields, while Fig. 6.7 shows the percentage of accepted patents.

6.4   University of Pennsylvania—USA

The University of Pennsylvania, a private university, managed to be listed in Reuters' ranking of the World's Most Innovative Universities thanks to its researchers [45]. It was established in 1740 and has been home to 4 Nobel Prize laureates and 4 National Medal of Science winners in the past two decades alone. The university comprises 137 research centers and institutes.

In 2016, the scientific research budget amounted to more than $1 billion in grants for sponsored projects. Also, in 2013, the University of Pennsylvania's business incubators and Penn Center for Innovation founded 22 start-ups that gained more than $50 million in revenue.

In 2017, the scientific research budget amounted to $928 million. Its scientists concluded tests on a new single-dose Zika vaccine and identified a targeted gene responsible for allowing host cells to defeat the Ebola virus. Researchers in Penn's Department of Biology identified a target gene that regulates the creation of hair-like root structures in plants which increase surface area, causing greater absorption of water and other resources from the soil. Engineered crop plants with this feature became most likely to resist drought conditions.

The university's school of medicine receives significant support from the National Institute of Health NIH in America, $10 million of which is dedicated to the establishment of the Physical Sciences Oncology Center at Penn, which will focus on the early detection and treatment of liver cancer.

Table 6.4 displays some key statistical figures that illustrate the quality of research and innovation [45–47].

Figure 6.8 shows the patents submitted by the University of Pennsylvania in various fields, while Fig. 6.9 shows the percentage of accepted patents.

Table 6.4 Some key statistical figures for the University of Pennsylvania

Number of Nobel laureates among professors and alumni	30
Endowments in 2017	$12.2 billion
University budget in 2017	$8.78 billion
Total campus area	4.39 km^2
Number of applicants for university (2017)	40,413
Number of students admitted to university (2017)	3757 (9.3% of applicants)
Number of undergraduate students	10,496 (48.8% of total students)
Number of postgraduate students	11,013 (51.2% of total students)
Teaching staff	4300
Members of the academic council	5242
Total number of university staff	16,000

Fig. 6.8 Patents submitted by the University of Pennsylvania

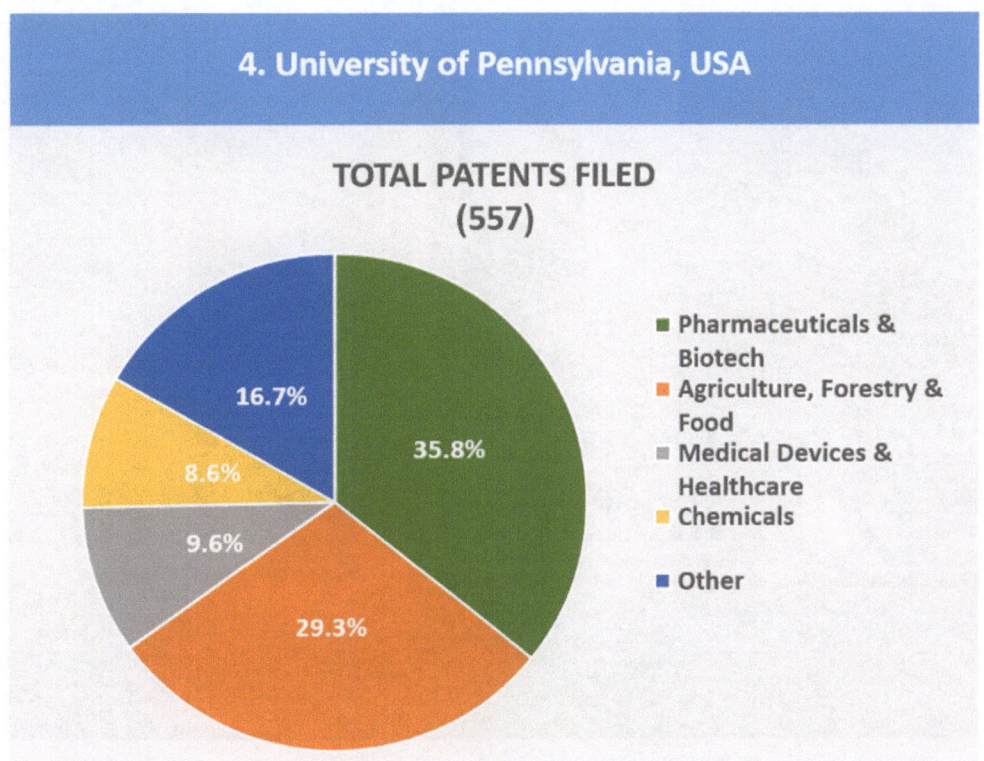

Fig. 6.9 Percentage of accepted patents of the University of Pennsylvania

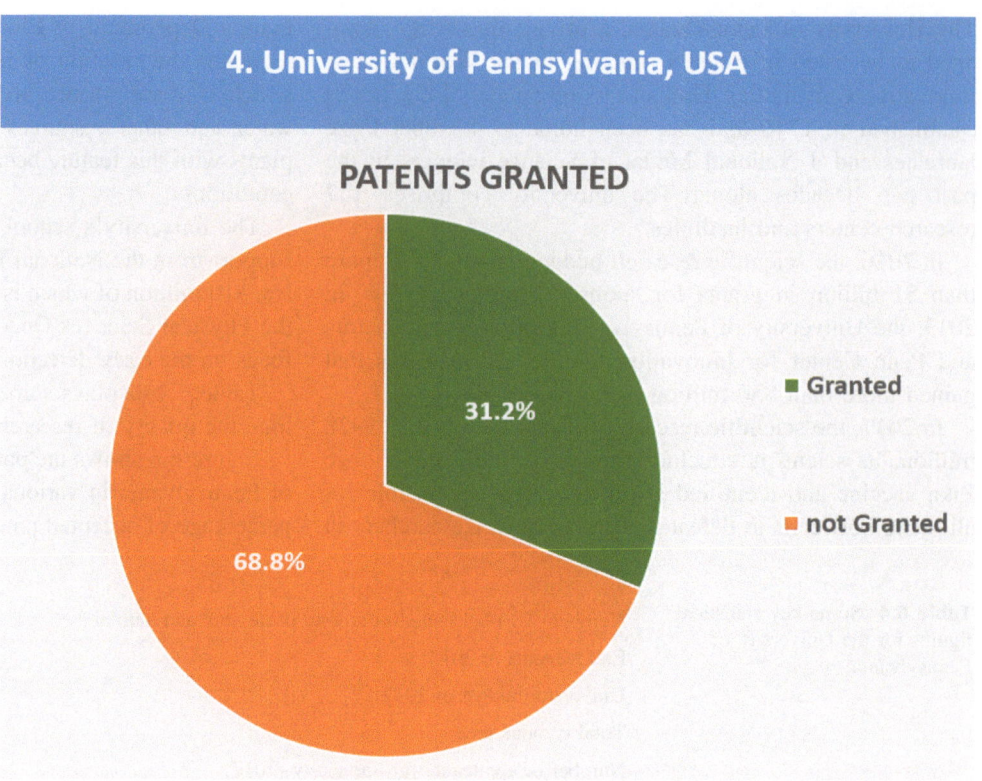

6.5   University of Washington—USA

Despite coming fifth in ranking, the University of Washington (UW) is ranked first among public universities on the list. It was established in 1861 and has three campuses in Seattle, Tacoma, and Bothell.

In the 2016 fiscal year, the research budget amounted to $995 million, which is 72% of the university's annual budget. That same year 21 new start-ups were launched making a total of 126 start-ups over the past ten years.

Materials engineering researchers at UW recently developed a fast and affordable way to create super capacitors that can store and discharge energy faster than conventional batteries. These supercapacitors use a low-density carbon material called aerogel, and there are applications for using them in a wide range of devices from electric cars to high-powered lasers. Moreover, researchers at the university have discovered that adding polydopamine (a chemical substance extracted from mussels) to various medical examinations increases examination accuracy a thousand times which could lead to better tests for HIV, Zika virus, and proteins related to cancerous tumors.

Table 6.5 displays some key statistical figures that illustrate the quality of research and innovation [48–49].

Figure 6.10 shows the patents submitted by the University of Washington in various fields, while Fig. 6.11 shows the percentage of accepted patents.

Table 6.5 Some key statistical figures for the University of Washington

Number of Nobel laureates among professors and alumni	7
Number of Fields Medalists among professors or alumni	1
Number of National Medal of Science winners	7
Number of Pulitzer Prize winners	2
Endowments in 2017	$3361 billion
Campus area	2.8 km^2
Number of undergraduate students (2017)	31,843 (67.5% of total students)
Number of postgraduate students (2017)	14,843 (32.5% of total students)
Members of the academic council	5803
Number of the university's administrative staff	16,174

Fig. 6.10 Patents submitted by the University of Washington

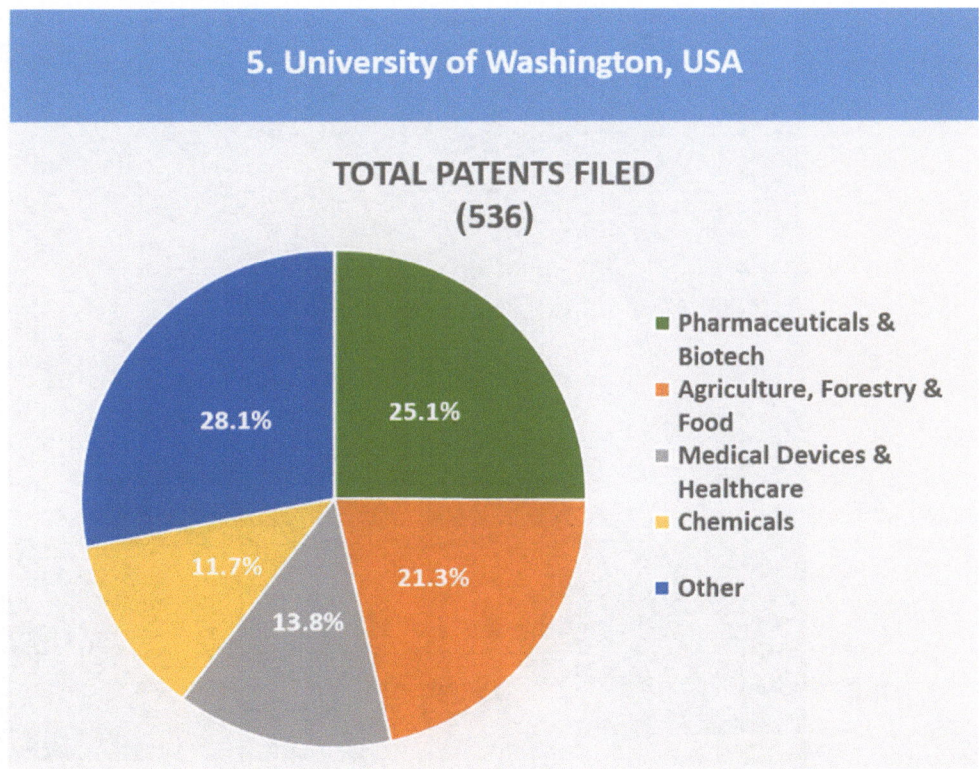

Fig. 6.11 Percentage of accepted patents of the University of Washington

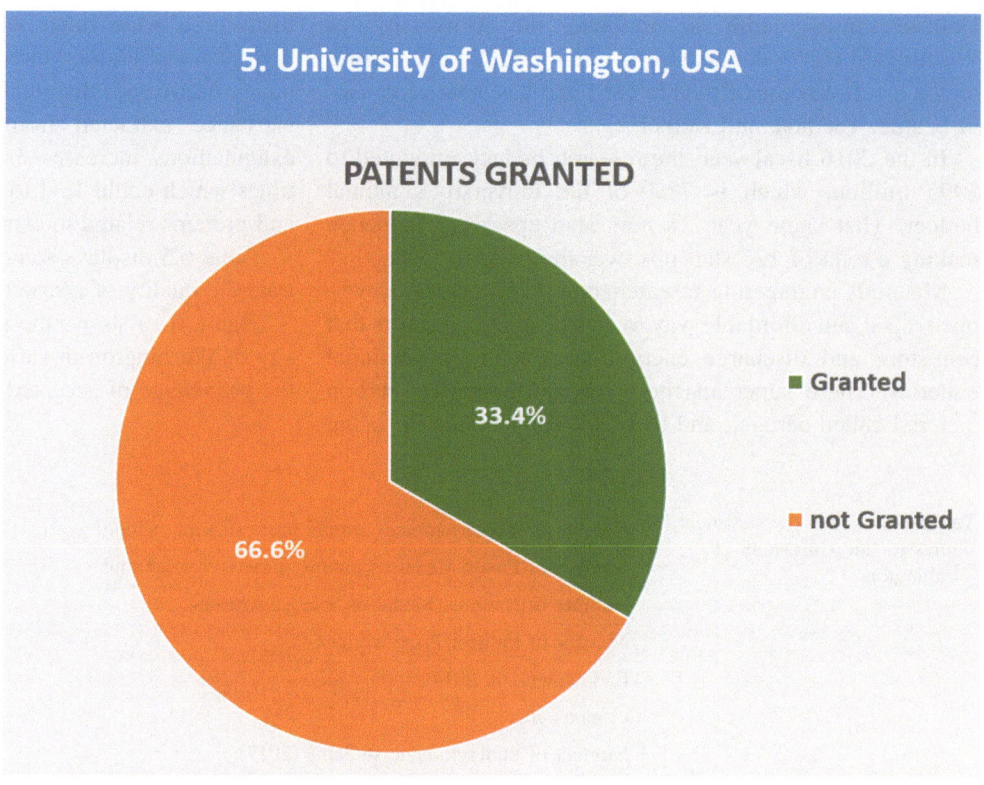

6.6   University of Texas System—USA

The eight public universities and six health institutions in the University of Texas (UT) System serve more than 221 thousand students each year, confer more than one-third of the state's undergraduate degrees, and educate two-thirds of the state's healthcare professionals annually. The system's faculty include 8 Nobel laureates. Its headquarters is in Austin, and the flagship campus there is the largest in the system.

This system receives almost 70% of all research funds dedicated to public institutions in Texas, and its annual research expenditures exceed $2.7 billion, of which $641 million is from private sector sources. Over the past five years, researchers from 14 institutions thereof have produced more than 120 thousand scientific papers.

In 2017, the University of Texas at Arlington announced the establishment of the Conrad Greer Lab, which will develop new methods of compressing natural gas into diesel and jet fuel. In the same year, the UT Austin campus built Stampede2, the most powerful supercomputer at any US university. As for the UT at Dallas campus, researchers invented a non-invasive biomonitor for diabetics, which uses the glucose levels in sweat to track blood-sugar levels.

Table 6.6 displays some key statistical figures that illustrate the quality of research and innovation [50–51].

Figure 6.12 shows the patents submitted by the University of Texas System in various fields, while Fig. 6.13 shows the percentage of accepted patents.

Table 6.6 Some key statistical figures of the University of Texas System

Endowments in 2017	$24 billion
Number of undergraduate students (2017)	167,028 (75.4% of total students)
Number of postgraduate students (2017)	54,309 (24.6% of total students)
Members of the academic council	17,158
Number of administrative staff	62,982

Fig. 6.12 Patents submitted by the University of Texas System

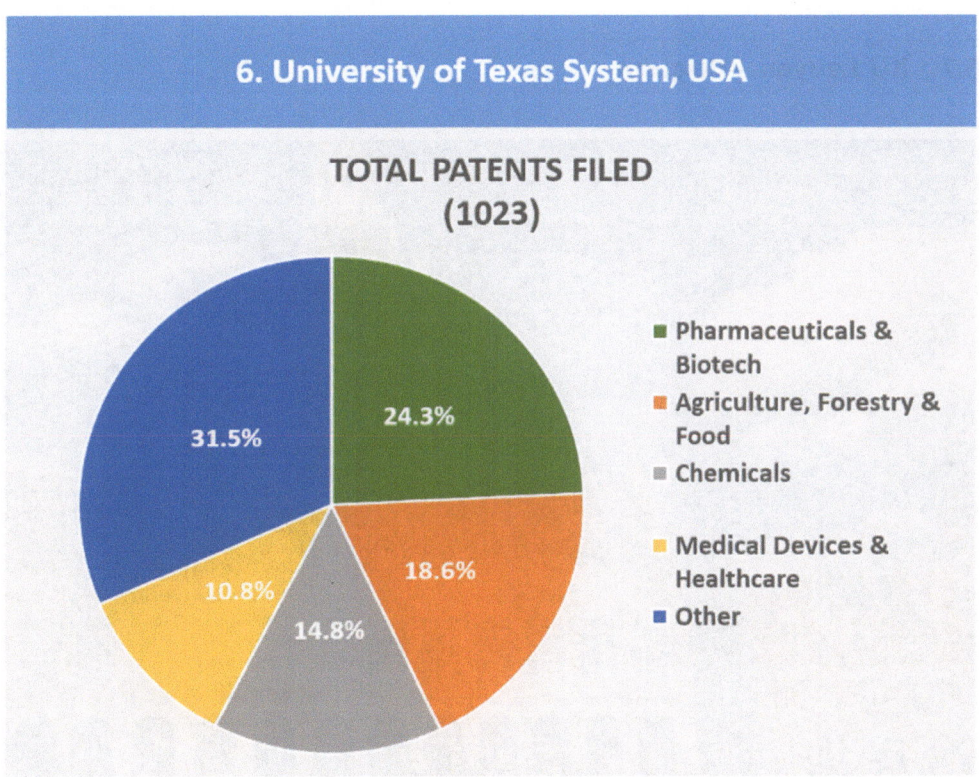

Fig. 6.13 Percentage of accepted patents for the University of Texas System

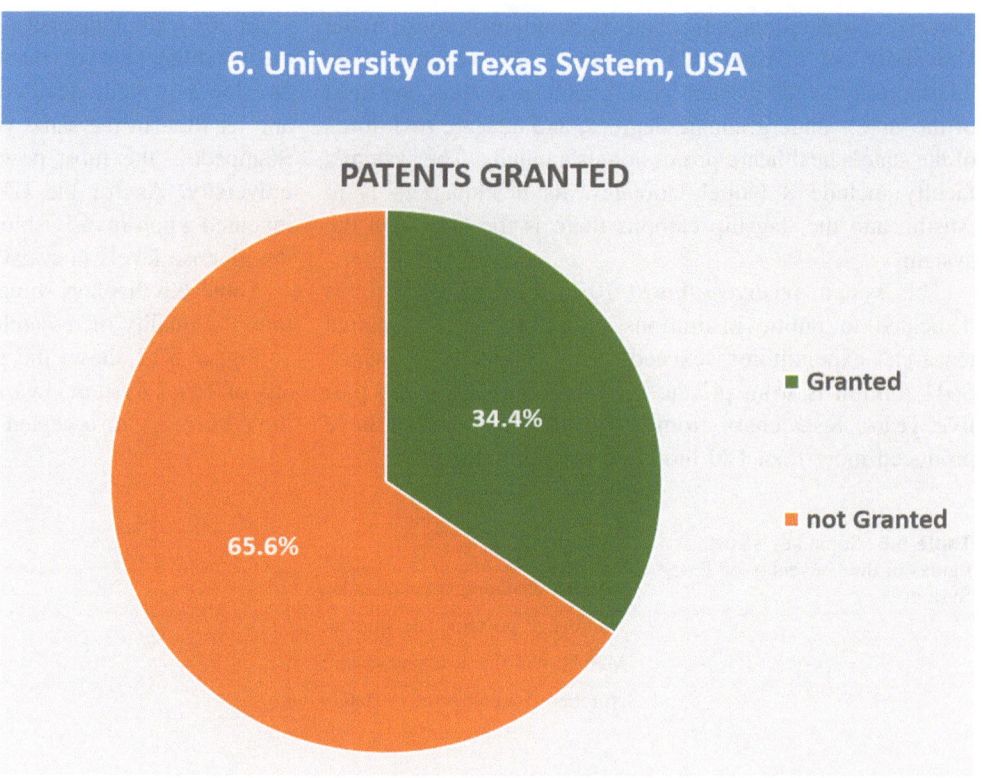

6.7   KU Leuven—Belgium

Founded in 1425 by Pope Martin V, KU Leuven is the oldest Catholic university, but it is not limited to theology. Part of KU's modern mission is to conduct comprehensive and advanced scientific research. A Dutch-speaking school based in Belgium's Flanders region, it is open to students of all faiths, operates independently from the Church and maintains one of the largest research and development organizations in the world [51].

KU Leuven Research and Development (LRD), established in 1972, was one of the first tech transfer offices in Europe, and has helped the university spin off more than one company across a range of industries. In 2012, Siemens acquired KU spinoff LMS International, a leading provider of mechatronic simulation software, in a deal worth approximately €680 million.

In the fiscal year 2015, the university's total research expenditure exceeded €454 million. The school's patent portfolio includes 586 patents.

In 2016, KU Leuven announced a new partnership with Ford Motor Company to study the durability of auto parts created using a "3D printing" manufacturing process. And in the eye surgery department, eye surgeons at University Hospital Leuven recently performed the first robotic surgery on a patient with retinal vein occlusion. Using a needle about 0.03 mm diameter, the robot injected a drug into the patient's retinal vein (which is about the diameter of a human hair. Both the robot and needle were developed by KU Leuven.

Table 6.7 displays some key statistical figures that illustrate the quality of research and innovation [52–53].

Figure 6.14 shows the patents submitted by KU Leuven in various fields, while Fig. 6.15 shows the percentage of accepted patents.

Table 6.7 Some key statistical figures for KU Leuven

Endowments in 2017	€950 million
Number of students	56,351
Number of doctoral students	4272 (7.6% of total students)
Members of the academic council	1107
Total number of university staff	11,534

Fig. 6.14 Patents submitted by KU Leuven

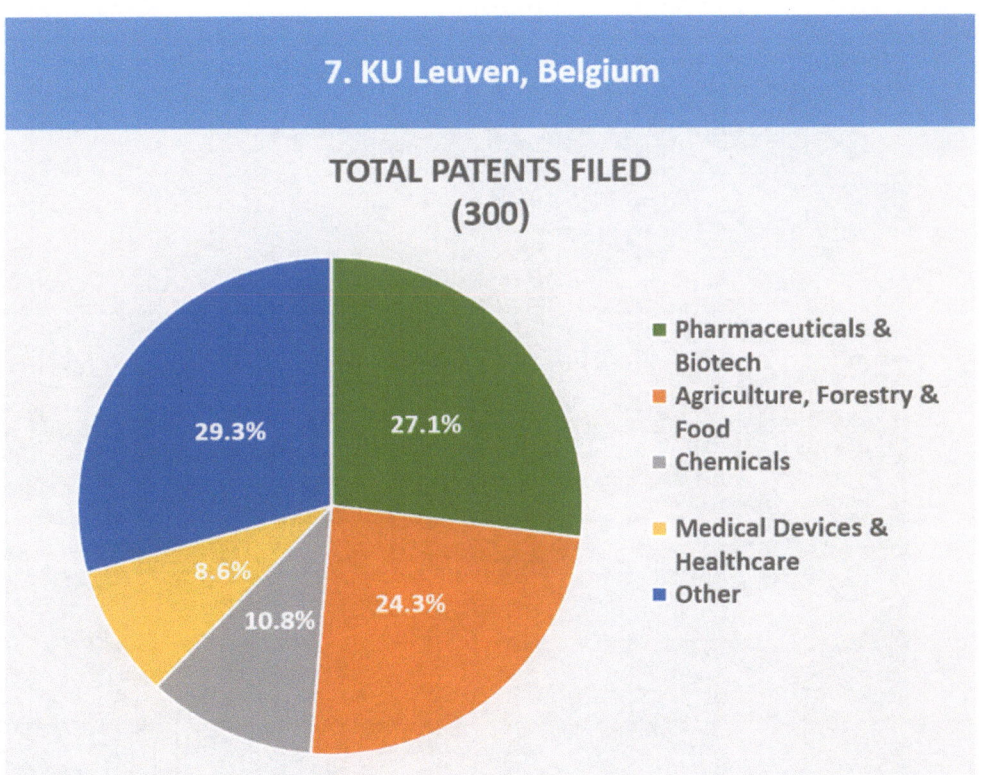

Fig. 6.15 Percentage of accepted patents of KU Leuven

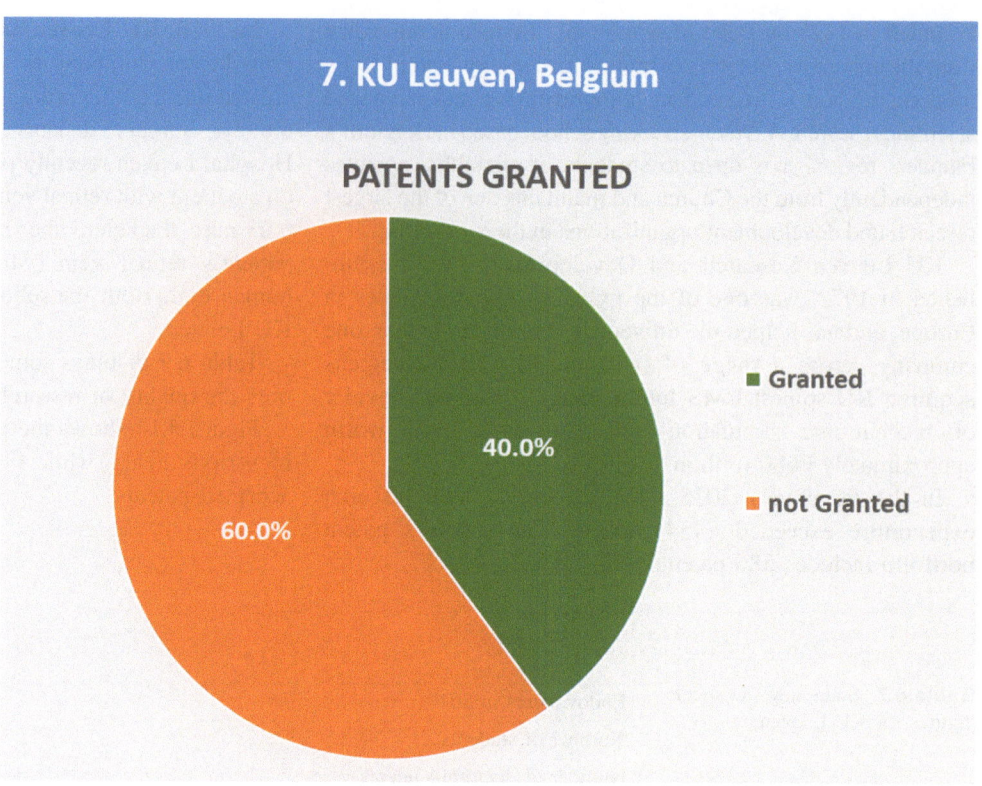

6.8   Imperial College London—United Kingdom

Imperial College London is a research university, founded in 1907 as a part of the University of London, and became fully independent in July 2007. It now focuses on four main disciplines: science, engineering, medicine and business. Throughout its history, Imperial College researchers have been responsible for innovations including the discovery of penicillin and the development of fiber optics. The university serves about 17,000 students from more than 125 countries and awards 6700 degrees annually. The university's 2015 annual research grant budget was about £436 million. This is in addition to the university's expenditures of research and equipment from its own budget, which amounted to £991 billion in 2017 [54].

Recent most important research at Imperial College London includes the development of nanoparticles that drive chemical reactions including artificial photosynthesis, which could be used in building more efficient solar panels. Imperial College scientists have also demonstrated a method for transferring power wirelessly to drones while hovering.

In 2016, the university opened a new innovation center called the I-Hub, which provides laboratories and office spaces to help entrepreneurs and start-ups commercialize new research. Imperial College's corporate partners include Shell and Qatar Petroleum, which fund the Carbon Storage Research Centre. Scientists there are investigating cleaner methods for the production and use of oil and gas.

Technology Company NEC is working with the university to make water systems more energy efficient and eco-friendly. Imperial College London is home to the Parkinson's UK Brain Bank, which supports more than 100 research projects and has more than 6000 registered potential tissue donors.

The 14 Nobel Prize laureates of Imperial College researchers and scientists enhance its excellence in innovation. Additionally, three young Imperial College mathematicians won the Fields Medal for mathematics, which is awarded to young mathematicians under the age of 40.

Table 6.8 displays some key statistical figures that illustrate the quality of research and innovation [54–57].

Figure 6.16 shows the patents submitted by Imperial College London in various fields, while Fig. 6.17 shows the percentage of accepted patents.

Table 6.8 Some key statistical figures of Imperial College London

Number of Nobel laureates among professors and graduates	14
Number of professors or graduates who are Fields Medal winners	3
Endowments (2017)	£141.7 million
University budget (2016/2017)	£983 million ($1.35 billion)
Annual grants value for research since 2013 until 2017	£330–460 million
Number of students	17,485
Number of undergraduate students	9503 (55% of total students)
Number of postgraduate students (2016)	7982 (45% of total students)
Acceptance rate of students	14.3%
Number of degrees awarded annually	6700
Members of the academic council	3765
Number of administrative staff	3940

Fig. 6.16 Patents submitted by Imperial College London

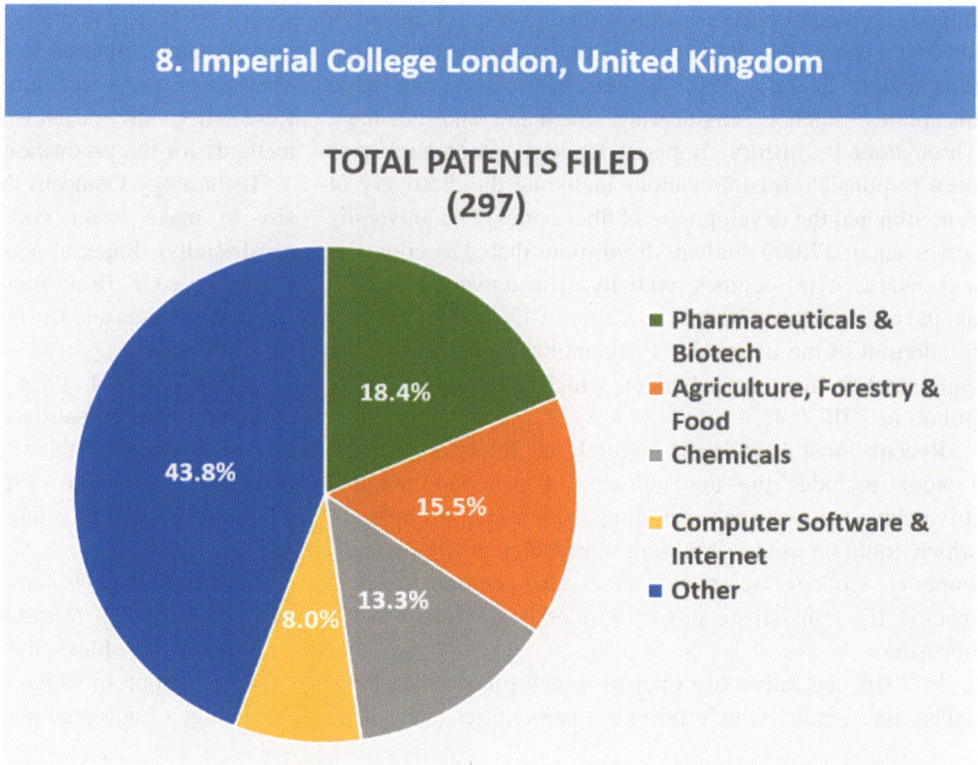

Fig. 6.17 Percentage of accepted patents of Imperial College London

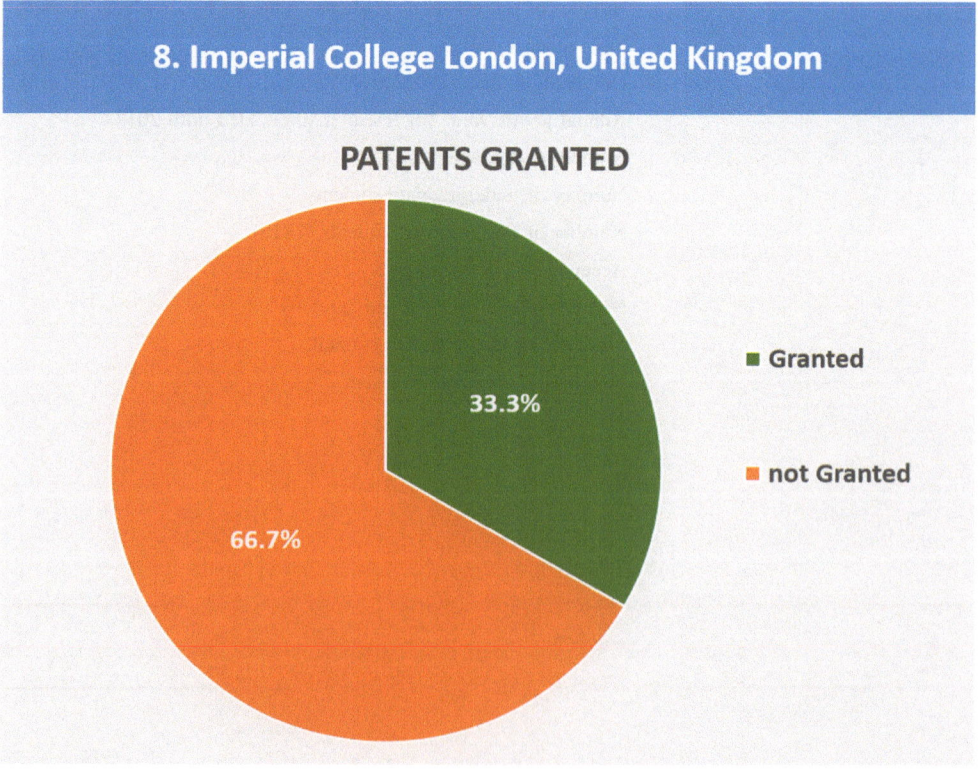

6.9 University of North Carolina Chapel Hill —USA

Table 6.9 Some key statistical figures of the University of North Carolina Chapel Hill

Number of Nobel laureates among professors and graduates	2
Campus area	729 acres (3 km^2)
Number of students (2017)	29,877
Number of undergraduate students (2017)	18,715
Number of postgraduate students (2017)	11,132
Teaching staff	3887
Number of administrative staff (2015)	8287

The University of North Carolina, a public research university in Chapel Hill, North Carolina, was founded on December, 1789. It is the flagship of the 17 universities of the University System of North Carolina. The university first began enrolling students in 1795, which makes it one of three schools to claim the title of the oldest public university in the USA. In 2015, US News and World Report ranked university of North Carolina fifth among the best US public colleges and universities.

In the past two decades, the university's research project expenditures have doubled, to the point that it gained reputation as a top national research university. The university's research expenditures exceeded $1 billion dollars during that period. Therefore, UNC-Chapel Hill is ranked sixth among research universities for federal funding devoted to development and research.

In 2016, the university conducted 767 research projects that directly tackled health, education and welfare of North Carolina citizens. The fact that it incorporated more than 3000 undergraduates who produced original research lays emphasis on its excellence in research.

Table 6.9 displays some key statistical figures that illustrate the quality of research and innovation [58–59].

Figure 6.18 shows the patents submitted by UNC-Chapel Hill in various fields, while Fig. 6.19 shows the percentage of accepted patents.

Fig. 6.18 Patents submitted by UNC-Chapel Hill

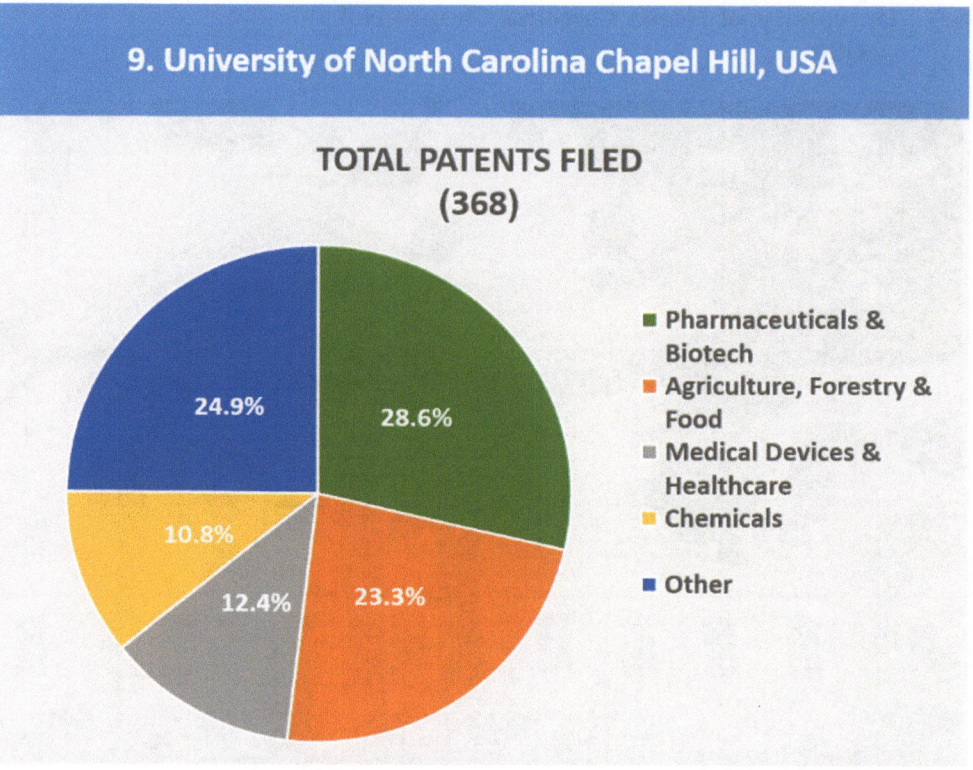

Fig. 6.19 Percentage of accepted patents for UNC-Chapel Hill

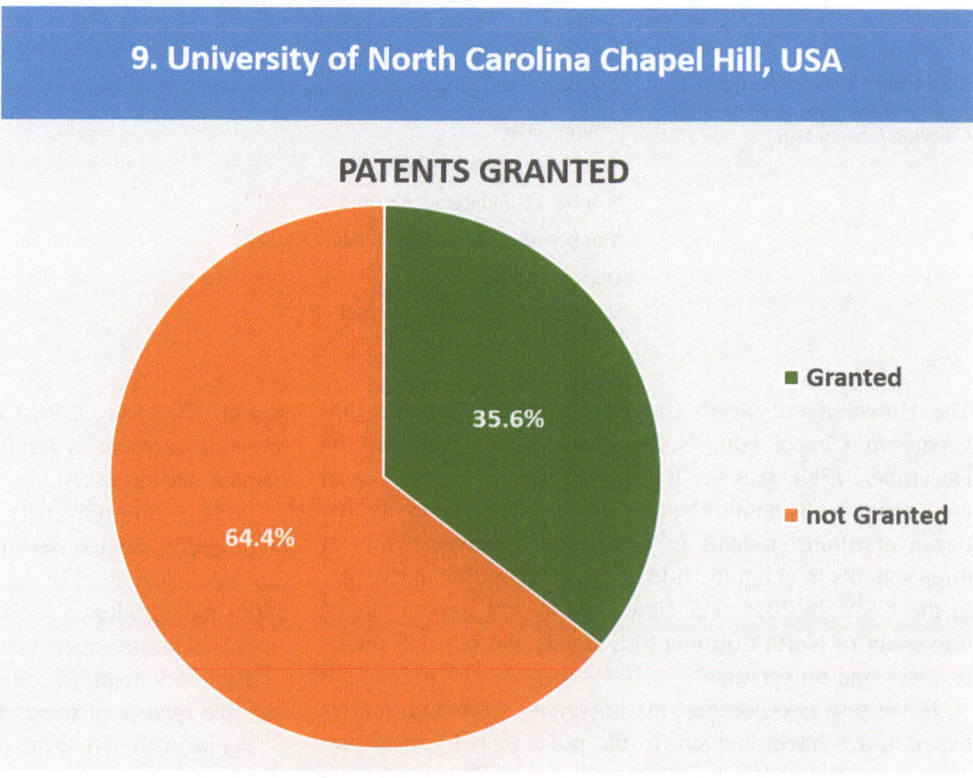

6.10 Vanderbilt University—USA

Table 6.10 Some key statistical figures for Vanderbilt University

Number of Nobel laureates among professors and alumni	3
Number of Fields Medalists among professors or alumni	2
Endowments in 2017	$4.1 billion
Campus area	1.3 km^2
Number of undergraduate students (2017)	6885 (54.7% of total students)
Number of postgraduate students (2017)	5707 (45.3% of total students)
Number of teaching staff members	1519
Members of the academic council	4102

Vanderbilt University was ranked in the tozp ten for the first time in 2017, climbing from No. 20 in 2016 as its patent citations by other researchers around the globe increased.

Vanderbilt University was founded in 1873 with a $1 million gift Mr. Cornelius Vanderbilt. Up to this day, it is a private institution that includes 10 schools and approximately 120 centers and institutes of various disciplines.

In the fiscal year 2016, the university's total research expenditures were approximately $235 million, and it sponsored research and project awards totaled $214 million. That same year the Center of Technology Transfer and Commercialization reported $6.5 million in revenue from licensing.

Recently, a team of Vanderbilt University researchers developed "smart underwear" that employs a series of straps to remedy back pain. Vanderbilt researchers have also developed circuit boards embedded with "cotton candy" tangles of silver nanowire that dissolve when cooled. The circuit boards are designed for use in devices of temporary function, including medical devices implantable for a temporary end and can be gotten rid of when not needed by cooling human body. Vanderbilt-developed products that have recently entered the market including lower limb exoskeleton that enables people with spinal cord injuries to stand and walk. Also a line of energy-saving pneumatic directional control valves for use in factory work.

Table 6.10 displays some key statistical figures that illustrate the quality of research and innovation [60–61].

Figure 6.20 shows the patents submitted by Vanderbilt University in various fields, while Fig. 6.21 shows the percentage of accepted patents.

Fig. 6.20 Patents accepted by
Vanderbilt University

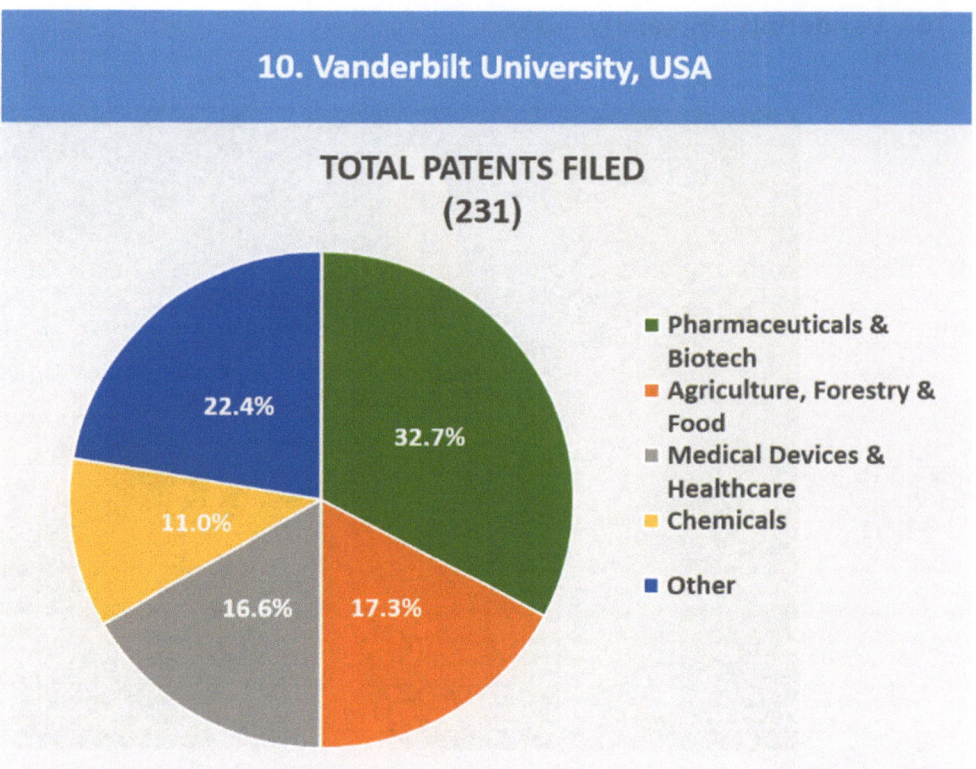

Fig. 6.21 Percentage of accepted
patents of Vanderbilt University

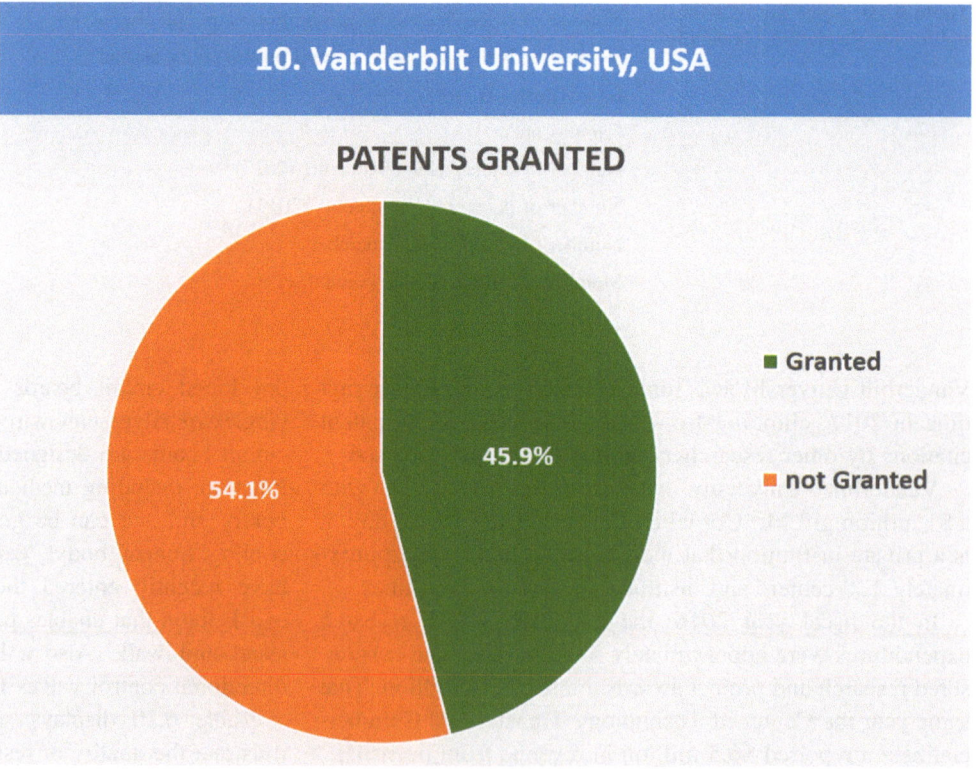

6.11   Korea Advanced Institute of Science and Technology (KAIST)—South Korea

Table 6.11 Some key statistical figures of KAIST University

Number of undergraduate students	4047 (39.5% of total students)
Number of postgraduate students	6202 (60.5% of total students)
Number of doctoral students	2311 (37.3% of postgraduate students)
Number of joint academic programs students	1137
Teaching Staff	1140

The full name of this university was Korea Advanced Institute of Science and Technology, but starting from 2008, only the abbreviation (KAIST) has been used [62].

KAIST is one of the oldest Korean research universities specializing in sciences and engineering. It operates in three cities; Daejeon, Seoul, and Busan. The university was established in 1971 by the Korean government on the model of US engineering faculties. At the beginning, it was funded with a loan worth millions of dollars by the United States Agency for International Development. Most curricula are delivered there in English, and the university still maintains strong relations with American academic institutions. The university supports joint academic programs, allowing students to get degrees from it as well as its partners such as Carnegie Mellon University and Georgia Institute of Technology.

The university provides students with a special entrepreneurship program called "Start-up KAIST," which helps to develop business initiatives and to provide financial support for new business. Yi So-yeon, the first Korean to fly in space, has graduated from such program.

In 2015, the cooperation between a university team and Rainbow, a company that emerged from (Humanoid Robot Research Center), a university research center, contributed to the university's being awarded by the US Department of Defense in a man-like adaptable robotics competition.

The university researchers have recently developed the "parasitic" robot system that controls the movement of the host animal. The robot is mounted on the carapace of the turtle and uses lights and food to encourage the animal to move in the required direction.

Among other pieces of research was the development of shock absorbing touch sensors that can be used as a robotic membrane. The mobile applications of this technique help to report traffic violations, such as breaking traffic signals or taking a banned turn.

Table 6.11 shows some key statistical figures of high-quality research and innovations [63].

Figure 6.22 shows patent applications submitted by KAIST in different fields, while Fig. 6.23 shows the percentage of accepted patent applications.

Fig. 6.22 Patent applications submitted by KAIST University

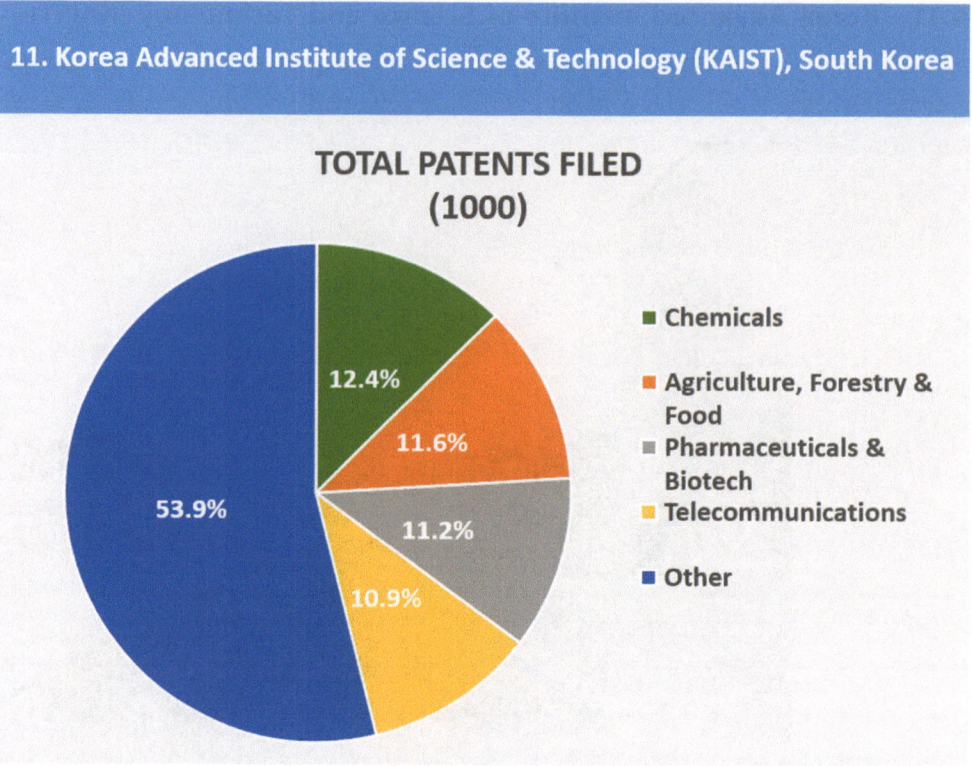

Fig. 6.23 Percentage of accepted patent applications submitted by KAIST University

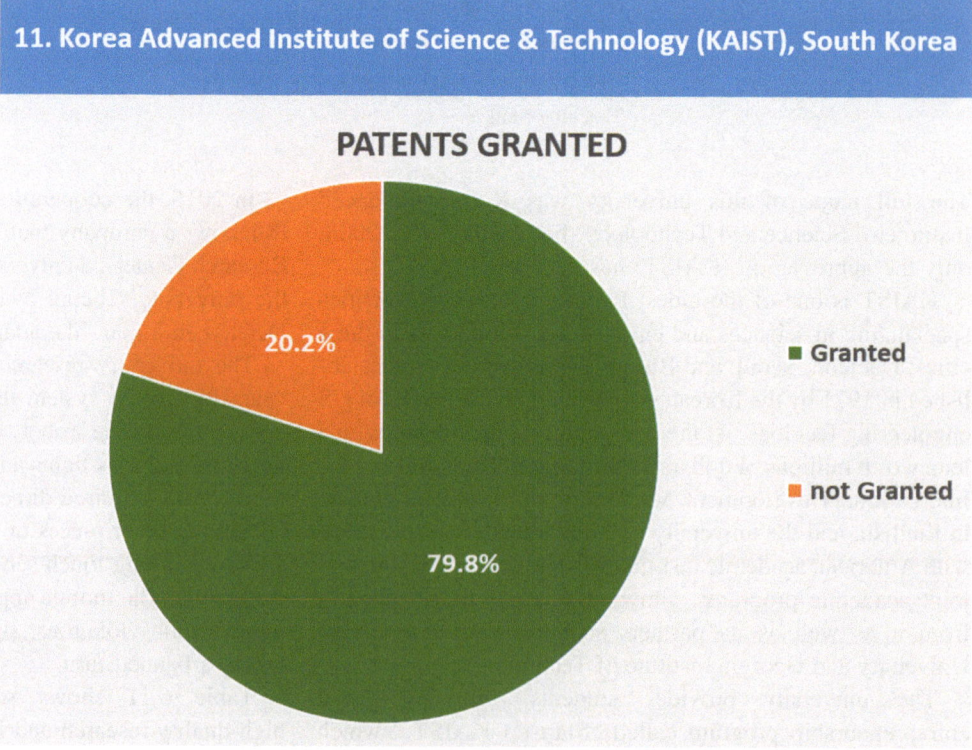

6.12   Swiss Federal Institute of Technology in Lausanne (EPFL)—Switzerland

Table 6.12 Some key statistical figures of Swiss Federal Institute of Technology in Lausanne

University budget (2017)	950 million francs (950 million U.S. dollars)
Number of students (2016)	10,343
Number of undergraduates (2016)	5418 (52% of total students)
Number of postgraduate students (2016)	4925 (48% of total students)
Number of teaching staff members	874
Number of participating professors	(27% of the total)
Number of assistant professors	(14% of the total)
Members of the academic council (2016)	3971
Number of administrative staff	1195
Number of start-ups	115

This institute is located alongside Lake Geneva and at the foot of the Alps. It is one of two Swiss federal institutes of technology, including more than 350 laboratories and research centers in the main campus. The institute also includes a massive computer used by several research groups. Among the projects carried out, there is the human mind simulation project, called Blue Brain Project. This project aims to provide neuroscientists with a better understanding of neurological diseases. The student–academic staff ratio is about 10:1, and the budget allocated for the institute is up to 950 million Swiss francs [64].

The institute has an innovation park, where laboratories, offices, and researchers from more than 150 companies and start-ups such as Nestlé, Siemens, and others are hosted. The start-ups associated with the innovation park have raised their budget to around 400 million Swiss francs in 2016, with an increase of 50% of 2014 total budget. Companies listed in 2016 include Mindmaze, a healthcare company, that develops virtual reality rehabilitation programs, with a budget of more than 100 million Swiss francs, i.e., the largest single amount reached by an unlisted company on campus. Another start-up, AC Immune, which develops a vaccine against Alzheimer's, has raised its budget to 42.7 million Swiss francs.

The institute has the Vice President Innovation (VPI) agency, which mainly links between the institute and business world, and encourages entrepreneurship and building start-ups. Industry helps develop partnerships with the Institute and fosters innovation and research. In 2017, the Institute initiated 15 start-ups, in addition to 115 companies in the Innovation Park. Such start-ups have raised their budgets to 112 million Swiss francs.

Table 6.12 displays some key statistical figures that illustrate the quality of research and innovation [64–66].

Figure 6.24 shows patent applications submitted by the Swiss Federal Institute of Technology in Lausanne in different fields, while Fig. 6.25 shows the percentage of accepted patent applications.

Fig. 6.24 Patents applications submitted by Swiss Federal Institute of Technology in Lausanne

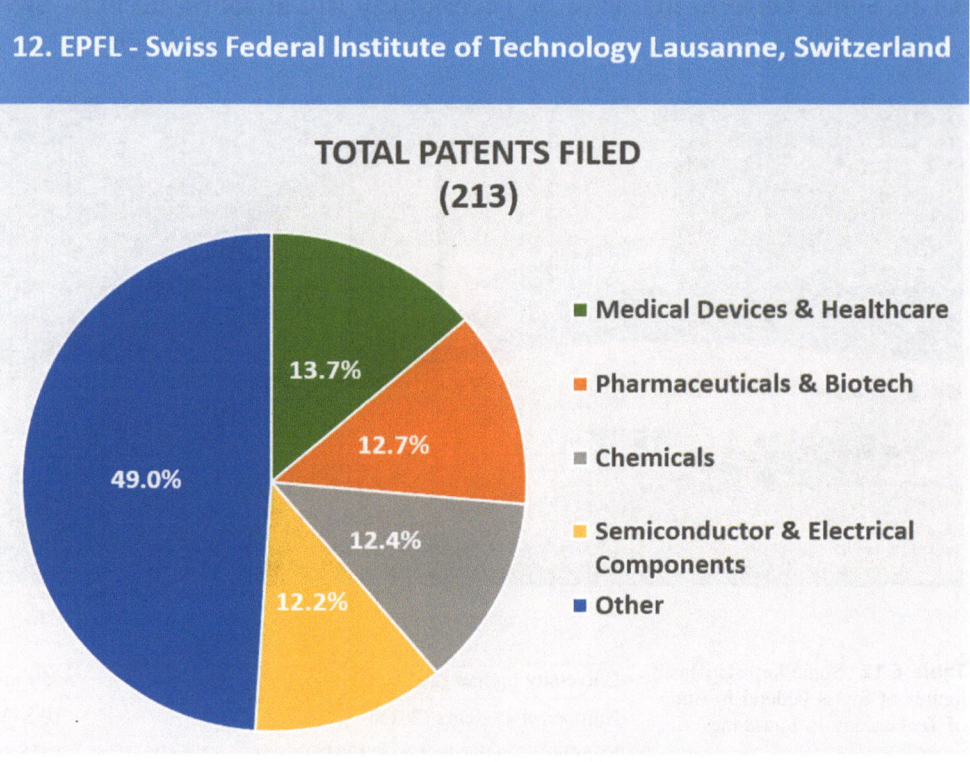

Fig. 6.25 Percentage of accepted patent applications of Swiss Federal Institute of Technology in Lausanne

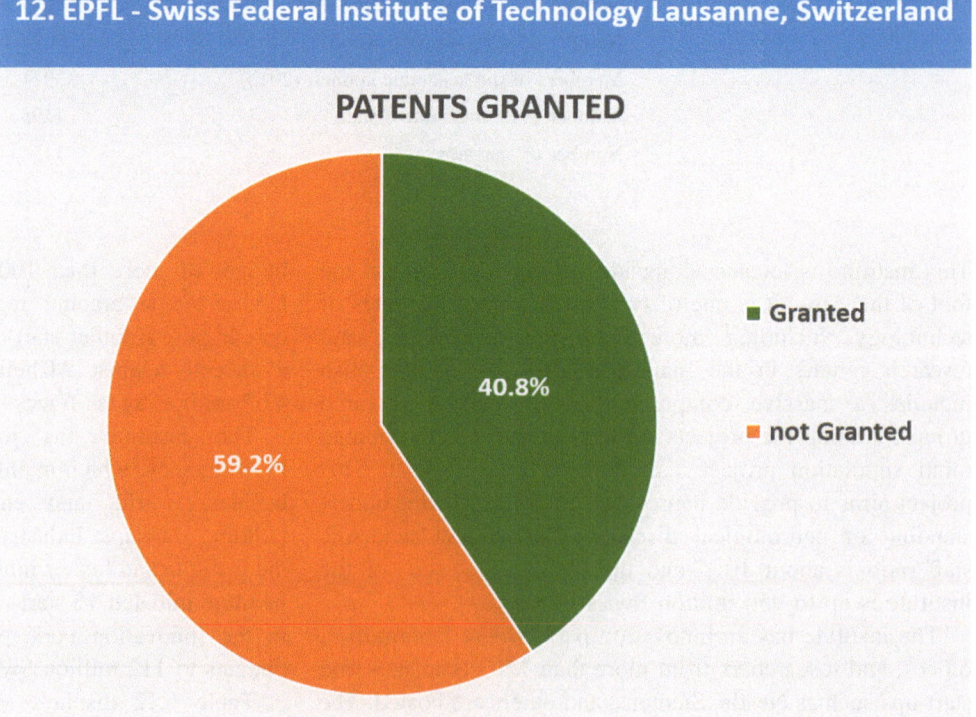

6.13 Pohang University of Science and Technology—South Korea

Table 6.13 Some key statistical figures of Pohang University of Science and Technology (POSTECH)		
The amount of scholarships and research contract (2016)		$155 million
The presented patents (2016)		479
The registered patents (2016)		372
Number of undergraduates (2016)		1449 (40% of total students)
Number of postgraduate students (2016)		2139 (60% of total students)
Number of teaching staff members		303
Members of the academic council		630

Pohang University of Science and Technology (POSTECH) was established in 1986 as a private research university closely related to the industry. Its 400-acre campus is just a few minutes away from the South Korean steel-making company headquarters POSCO). Its revenues from the scholarships and research contracts exceeded $155 million in fiscal year (FY) 2016, accounted for 46% of the overall budget of the university. It is a high percentage leading to excellence in research and inventions.

The number of students in Pohang University is 3588, the number of teaching staff members is 303 (student–teacher ratio is 10:1), and the university has four departments of science and seven departments of engineering, i.e., only 11 scientific departments. The university has 72 research units, including an institute for CO_2 reduction, a robotics lab and a machine learning center. The university also has a radiation accelerator laboratory, the only one in South Korea [67].

In 2016, the university developed X-Ray Free-Electron Laser (XFEL) instrument, an instrument that will allow scientists to observe the material at the nanoscale level and track the movements at ten trillionths of a second. This instrument is the third one of its kind in the world. Modern innovations include the development of an economical and environment friendly way of creating waterproof surfaces, namely through coating surfaces with salt granules (particles) being commercially available.

Important facts leading to distinctiveness in innovation include the fact that postgraduate students reached (2139) in 2016G, exceeding the number of undergraduate students of (1449) and accounting for 60% of the total number of students. With respect to researchers, their number reached 630 researchers, i.e., double the number of teaching staff members of 303.

Table 6.13 displays some key statistical figures that illustrate the quality of research and innovation [67–68].

Figure 6.26 shows the patent applications submitted by POSTECH in different fields, while Fig. 6.27 shows the percentage of accepted patent applications.

Fig. 6.26 Patent applications submitted by Pohang University of Science and Technology (POSTECH)

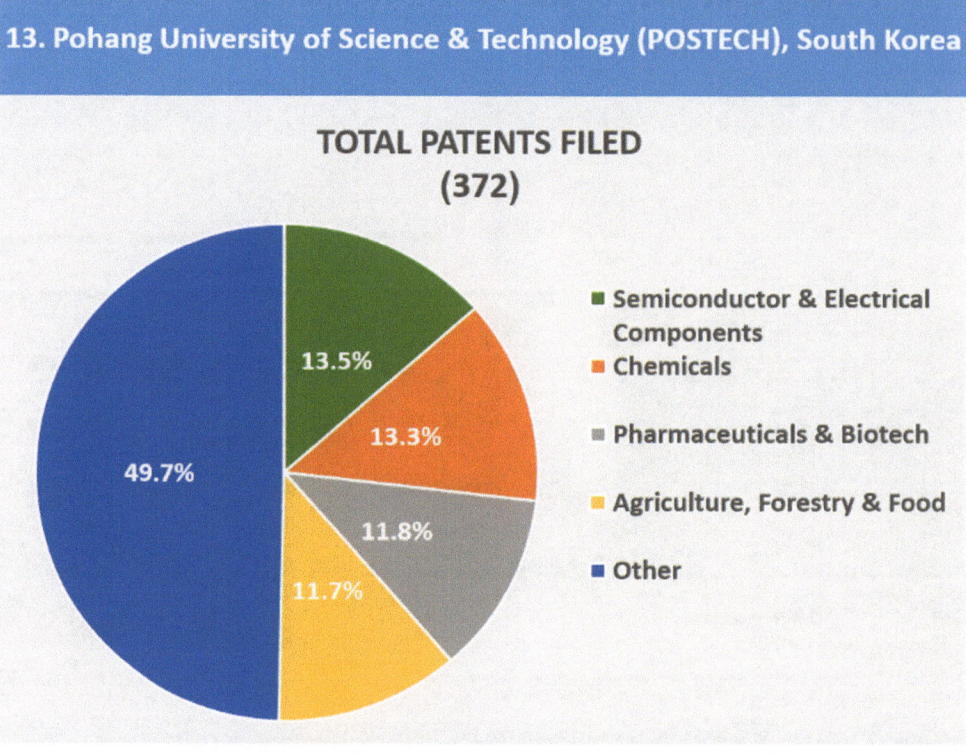

Fig. 6.27 Percentage of accepted patent applications of Pohang University of Science and Technology (POSTECH)

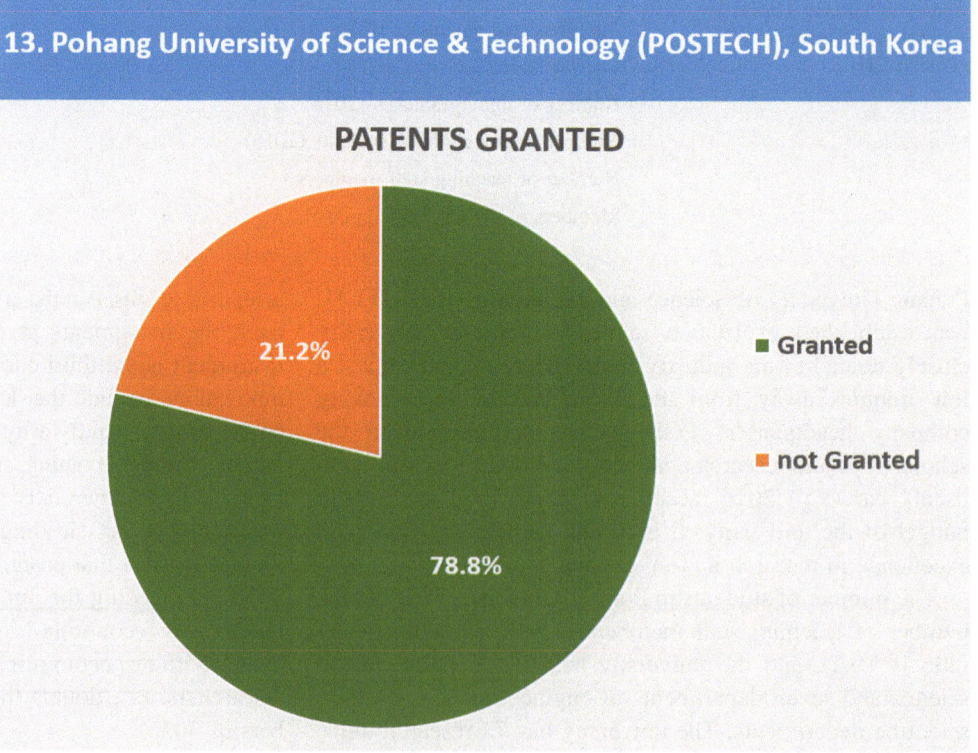

6.14 University of California System—USA

In 1869, University of California, a state public university located in California, started with only 10 teaching staff members and 38 students. Today, it includes 10 campuses (located in Berkeley, Davis, Irvine, Los Angeles, Merced, Riverside, San Diego, San Francisco, Santa Barbara, and Santa Cruz) and serves more than 273,000 students, including 56,432 postgraduate students. The university employs more than 22,000 teaching staff members, with a total of 1,700,000 graduates who live and work around the world. The university system operates five medical centers, and since 1943, it has been involved in managing three national laboratories (Lawrence Berkeley, Lawrence Livermore, and Los Alamos) for the US Department of Energy [69].

University of California is ranked the first in terms of the total number of patent applications submitted in 2015. The ten affiliated universities produce five new inventions on average per day. The university has received more than $4.97 million to fund research conducted this year.

Recently, California researchers in Los Angeles have discovered a new way to activate stem cells in hair follicles to help hair grow. This discovery can lead to new medications that promote hair growth for people with alopecia or hair loss.

In July 2015, Defense Advanced Research Projects Agency (DARPA) provided $21.6 million to scientist in California Berkeley to develop a "brain modem" that can directly stimulate thousands of neurons using expected light. This research could lead to devices that replace damaged eye by concentrating light and directly addressing the visible cerebral cortex.

Additionally, the university has more than 160 academic specializations, 600 postgraduate programs, 61 Nobel laureates, and 20,000 courses. This has contributed with $46.3 billion to California's economy. Until February 2011, 1029 start-ups have been established and 12,450 patents have been activated. The operating of 2017/2018 is approximately $34.5 billion.

Table 6.14 displays some key statistical figures that illustrate the quality of research and innovation [69–71].

Figure 6.28 shows the patent applications submitted by the University of California System in different fields, while Fig. 6.29 shows the percentage of accepted patent applications.

Table 6.14 Some key statistical figures of University of California System

Number of Nobel laureates among professors and graduates	61
Number MacArthur Genius' Grant Winners among professors and graduates	90
Number of National Medal among professors and graduates	67
Number of university campuses	10
Contractual Research Scholarships of Federal Government (2017)	$6500 billion
University budget (2017/2018)	$34.5 billion
Returns of university activities	$46.3 billion
Number of patents (2017)	1803
Number of undergraduates (2016)	216,568

(continued)

Table 6.14 (continued)

Number of postgraduate students (2016)	56,432
Teaching staff	22,700
Members of the academic council	45,700
Number of administrative staff	54,900

Fig. 6.28 Patent applications submitted by University of California System

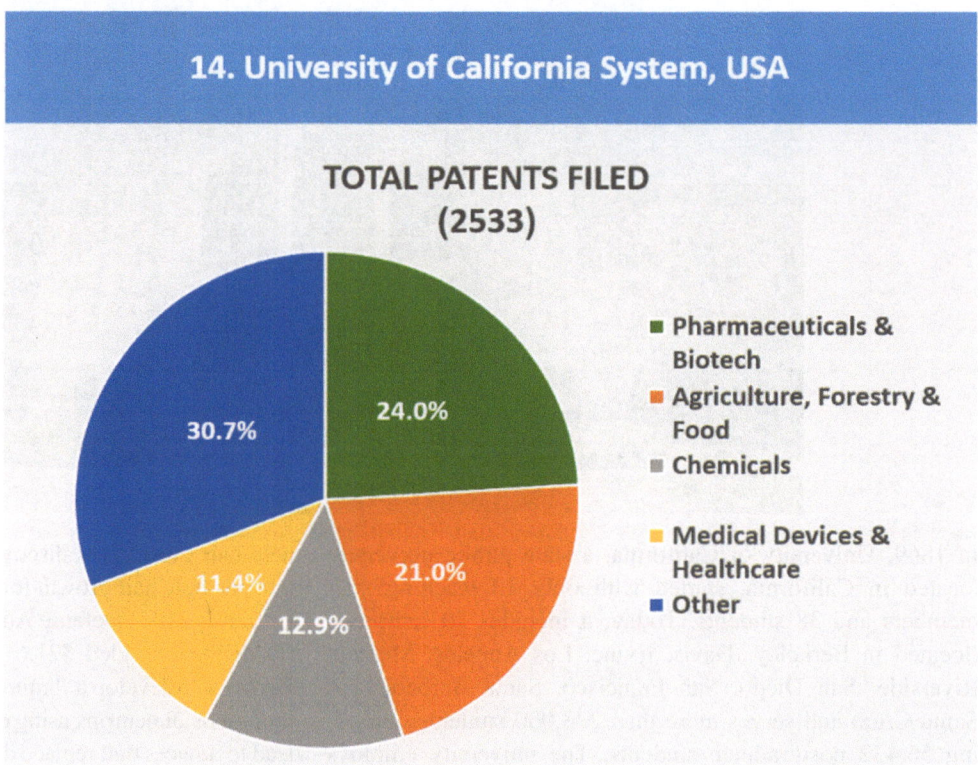

Fig. 6.29 Percentage of accepted patent applications submitted by University of California System

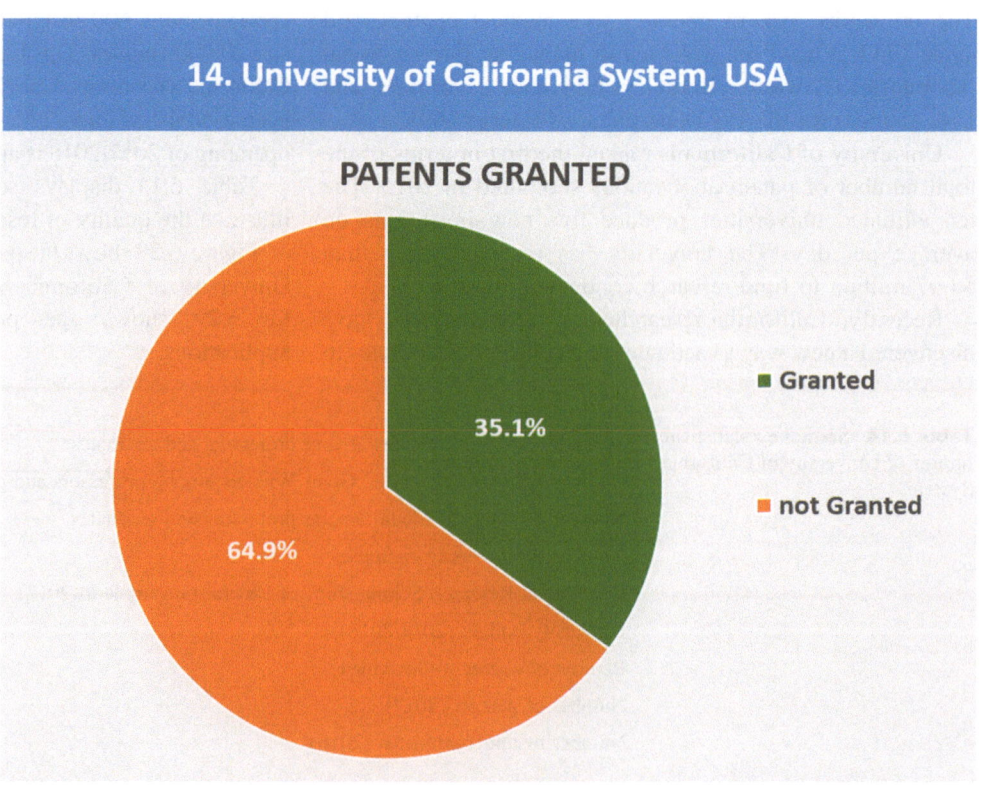

6.15 University of Southern California—USA

A private research university located in Los Angeles, University of Southern California, maintains strong relations with Hollywood and boasts great experience in digital entertainment. It also has the largest research program in computer science compared to all American universities, as well as the largest engineering and health programs compared to private universities. The University Institute of Innovative Technologies teaches how people interact with technology through virtual characters and simulations. The university also includes the Digital Arts Center, the only center in the USA specializing in digital filmmaking. The student–academic staff ratio is 10:1 and the percentage of accepted compared to applicant students in 2017 is 16% [72].

In a study of lifestyle and health history of more than 185,000 people, researchers from Keck School of Medicine of the University of Southern California have found that coffee drinkers live longer, finding that drinking coffee is associated with a lower risk of dying from heart disease, cancer, stroke, and diabetes, respiratory and kidney diseases.

Stevens Center for Innovation, the University's Technology Transfer Center, has initiated a number of start-ups in

Table 6.15 Some key statistical figures of the University of Southern California

Endowments (2017)	$5.1 billion
University budget (2017)	$4.9 billion
Research support budget (2017)	$0.76 billion
Returns of university activities	$8.0 billion
Campus area	308 acres (1.25 km^2)
Number of students (2017)	45,871
Percentage of students admitted to the university (2017)	16% of applicants
Number of undergraduate students (2016)	19,371 (42% of total students)
Number of postgraduate students (2016)	26,500 (58% of total students)
Members of the academic council	4361

the entertainment industry. The university budget for 2017/2018 increased around $4.9 billion, including $764 million for research support. Endowment funds reached $5.1 billion at the end of June 2017, while the returns of university activities amounted to $8 billion.

Table 6.15 displays some key statistical figures that illustrate the quality of research and innovation [72–74].

Figure 6.30 shows the patent applications submitted by the University of Southern California in different fields, while Fig. 6.31 shows the percentage of accepted patent applications.

Fig. 6.30 Patent applications submitted by the University of Southern California

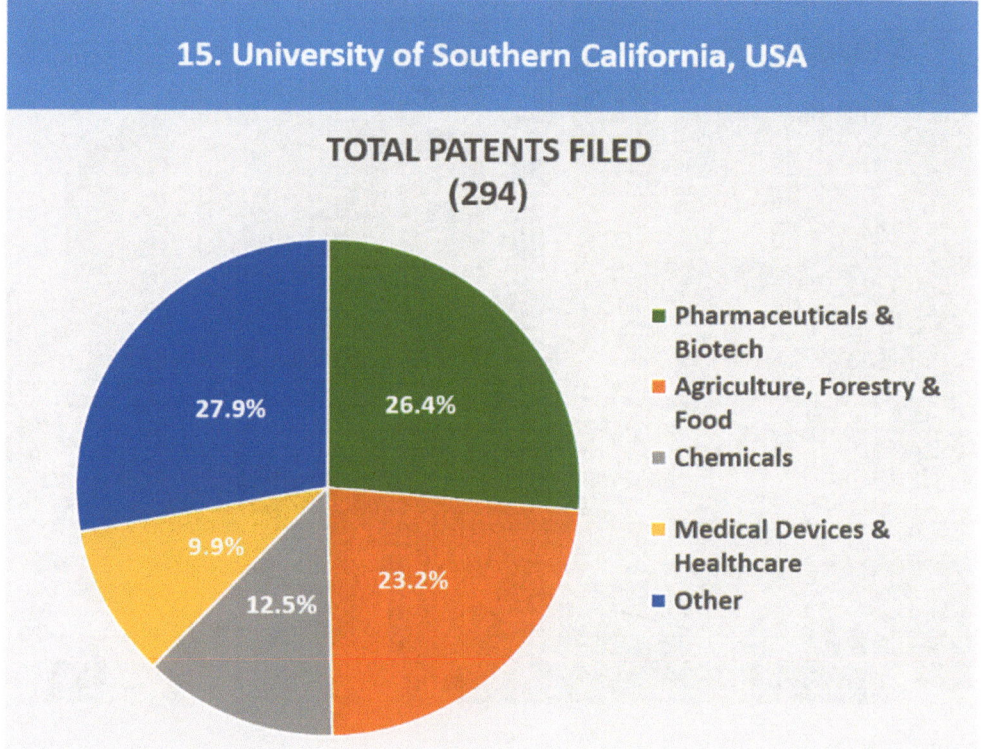

Fig. 6.31 Percentage of accepted patent applications submitted by the University of Southern California

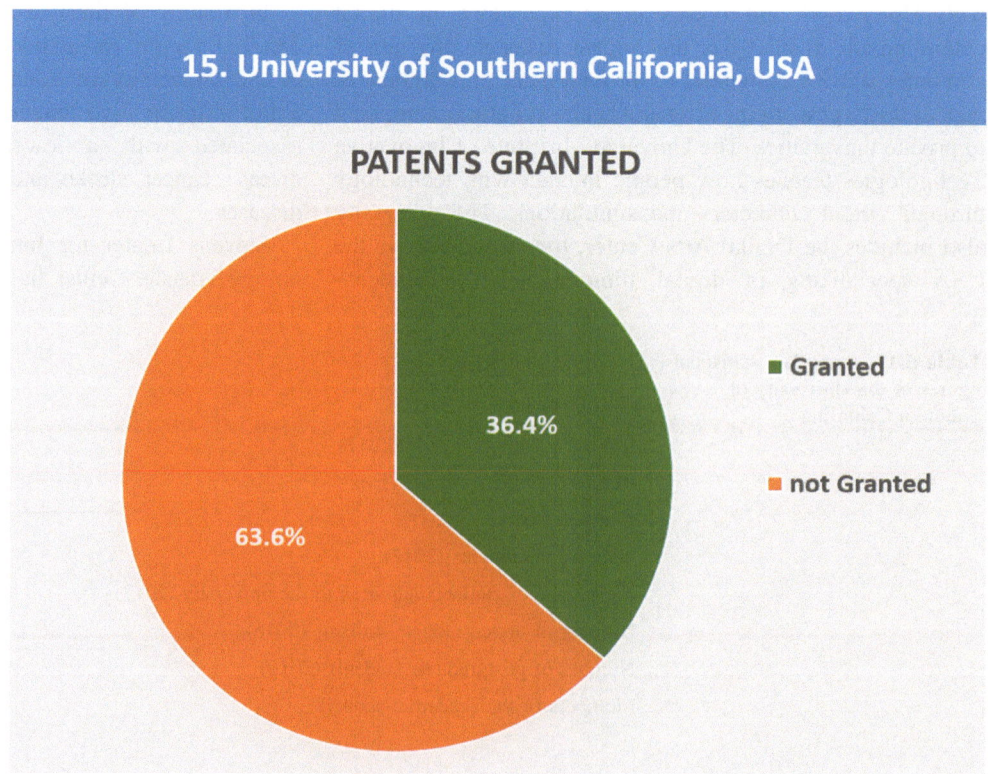

6.16 Cornell University—USA

Cornell University was founded in 1865, a private university with three university locations (Ithaca, New York and Doha, Qatar). The Ithaca campus has seven undergraduate colleges and four postgraduate specialized colleges, and the New York campus has two postgraduate medical schools. The number of students in the university is 22,319, of whom 5605 are in postgraduate studies. The academic staff-to-student ratio is estimated to be a member to about eight students (at the postgraduate and undergraduate levels), which helps the university to excel in scientific research and thus innovation [75].

In 2017, Cornell researchers used a high-energy X-ray scanner (called Cornell High Energy Synchrotron Source,

CHESS) to study and improve welding techniques for construction and engineering for Caterpillar Inc., in order to reduce harmful stresses on the materials resulting from the welding process. Researchers at the university have created a new cost-competitive process that uses polymer-coated nanofibers to clean contaminants from wastewater. Nanofibers form a surface for holding and spreading beneficial bacteria, eliminating contaminants.

Table 6.16 displays some key statistical figures that illustrate the quality of research and innovation [75–77].

Figure 6.32 shows patent applications submitted by Cornell University in different fields, while Fig. 6.33 shows the percentage of accepted patent applications.

Table 6.16 Some key statistical figures for Cornell University

Number of Nobel laureates among professors and graduates	3
Number of Crafoord Prize laureates (Crafoord Prize) among professors or alumni	1
Number of professors or graduates who are Fields Medal winners	1
Endowments (2017)	$6.8 billion
Number of university campuses	3
Average number of students admitted to the university annually	14.1% of applicants
Number of undergraduate students (2016)	16,714 (77% of total students)
Number of postgraduate students (2016)	5605 (23% of total students)
Number of teaching staff members	2908
Number of academic staff	2504
Number of administrative staff	8253

Fig. 6.32 Patents submitted by Cornell University

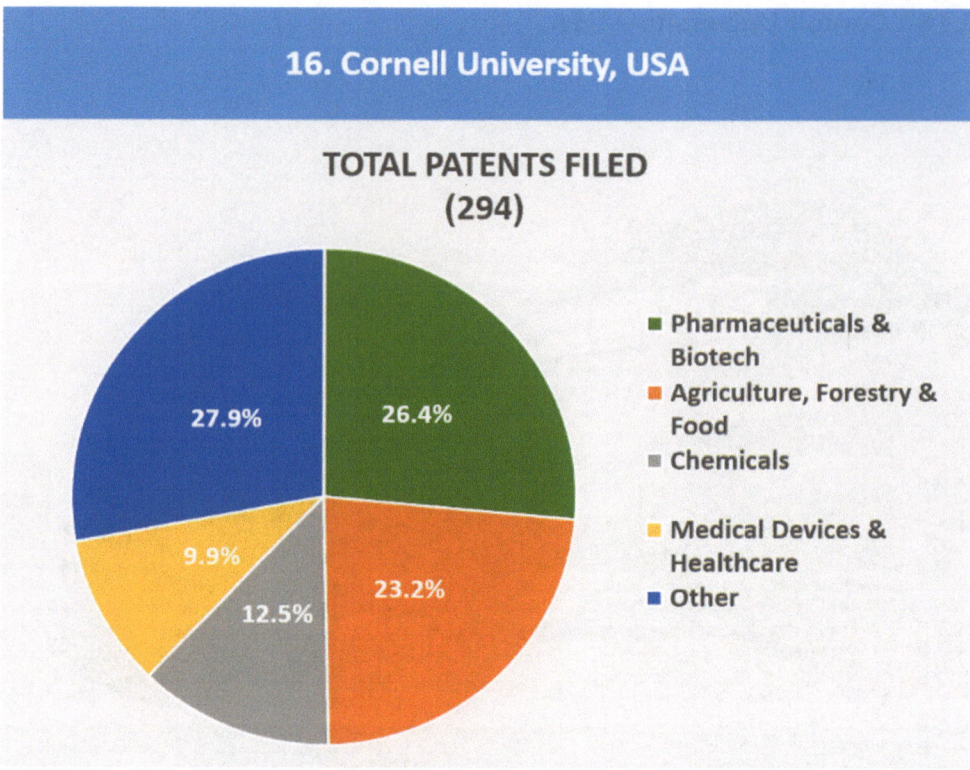

Fig. 6.33 Percentage of patents accepted for Cornell University

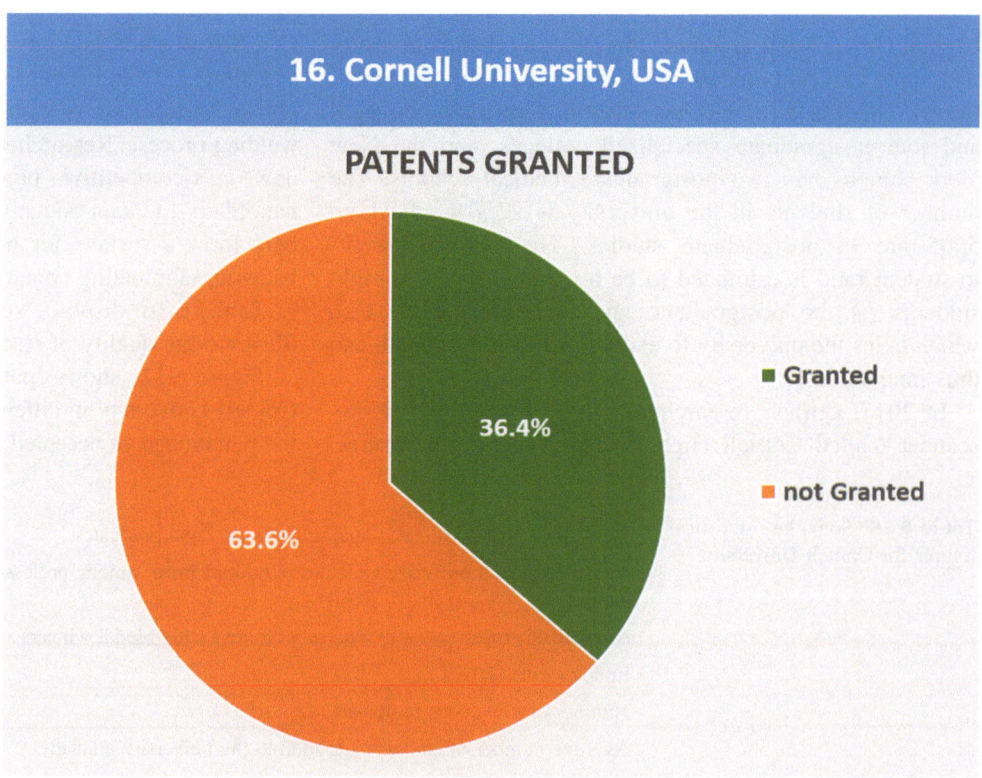

6.17 Duke University—USA

Duke University, the oldest American university, is located in the city of Durham in the state of North Carolina. It was founded by John Harvard in 1838 to be a counterpart to the British Universities of Cambridge and Oxford. It is a private research university that seeks, through the Innovation and Entrepreneurship Initiative, to conduct university research that results in actual, tangible products [78].

The University has many important centers and institutes, such as the marine science laboratory (Duke Marine), the cancer institute (Duke Cancer), and human vaccine institute (Duke Human Vaccine). Duke University played an important role in the creation of the most important social innovation of the twenty-first century. Its graduate, Soleio Cuervo, who graduated in 2003, created the famous "Like" feature, when he worked as a designer on social media site, Facebook.

The number of students at the university is 15,192, and the number of academic staff is 5022, which means that the student-to-teaching staff ratio is 3:1, making the interaction between students and the teaching staff very significant, and this is one of the important factors in the development of the spirit of innovation among students. In addition to that, the student admission rate did not exceed 9% of applicants in 2016, which means choosing the most outstanding students among those who have met the university's application requirements.

Researchers at Duke University have recently received a $7 million grant from The Bill and Melinda Gates Foundation, based on two previous grants totaling $10.7 million, to finance the "Reinvent the Toilet" project, which aims at developing a neutral, non-networked sewer system, as nearly 40% of the world's population suffer from the lack of proper sanitation facilities. As a result, approximately 525.000 children under the age of five die each year from diarrhea diseases, according to the World Health Organization. The university researchers will use the new financial support to test models of their system in India and South Africa.

Table 6.17 displays some key statistical figures that illustrate the quality of research and innovation [78–80].

Figure 6.34 shows patent applications submitted by Duke University in different fields, while Fig. 6.35 shows the percentage of accepted patent applications.

Table 6.17 Some key statistical figures for Duke University

Number of Nobel laureates among professors and graduates	11
Number of professors or graduates who are Turing Prize winners	3
Endowments (2017)	$7.9 billion
University budget (2017)	$2.3 billion
Number of applicants for university (2017)	33,110
Number of students admitted to university (2017)	1744 (5.3% of applicants)
Number of undergraduate students (2016)	6532 (43% of total students)
Number of postgraduate students (2016)	8660 (57% of total students)
Number of professors	988 (59% of all)
Number of participating professors	454 (27% of all)
Number of assistant professors	288 (14% of all)
Members of the academic council	5022
Number of administrative staff	8435

Fig. 6.34 Patents submitted by Duke University

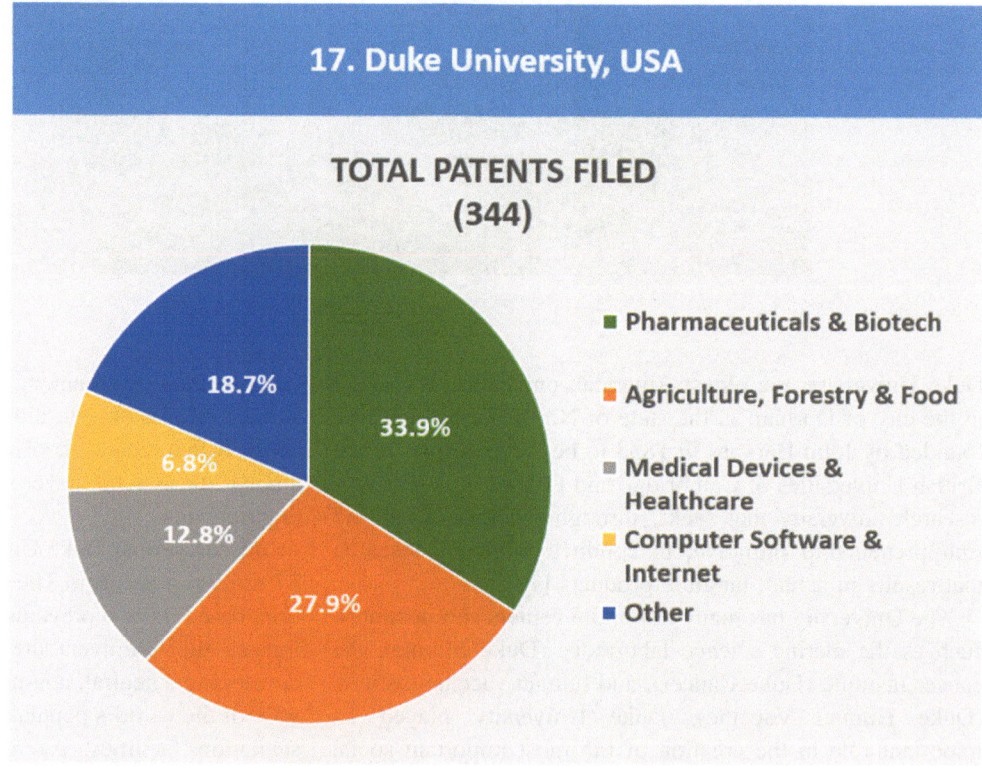

Fig. 6.35 Percentage of patents accepted for Duke University

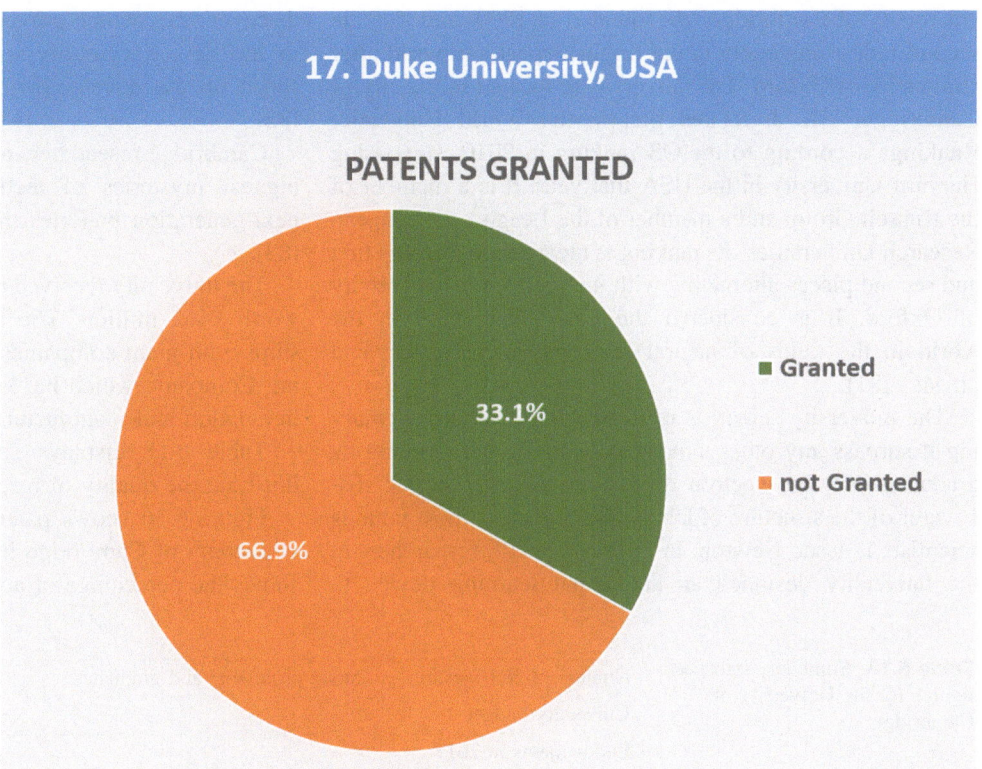

6.18 University of Cambridge—United Kingdom

University of Cambridge was founded in 1209, and it is the second oldest university in the English-speaking world after University of Oxford. The university is located in the city of Cambridge, UK. It ranked first in the World University Rankings according to the QS ranking in 2010, surpassing Harvard University in the USA that year. It is a member of the Russell Group and a member of the League of European Research Universities. Its ranking is always between the first and second places alternating with the prestigious University of Oxford. It is considered the oldest university in the world in the fields of natural sciences, mathematics, and physics [81].

The university scientists received 89 Nobel Prizes, making it surpass any other university in the world. Among its graduates are the electron discoverer, as well as the discoverer of the structure of DNA, and one of its most famous scientists is Isaac Newton. In February 2019, researchers at the university designed an automated learning device to discover drug, which proved to be doubly efficient compared to the device currently used in the industry. This could speed up the development of new treatments for diseases [82].

Cambridge researchers are working on solving one of the biggest mysteries of technology, that is, how to build next-generation batteries that can ignite a green revolution [83].

The university received research grants at the end of 2016 worth £462 million. The university has strategic relationships with giant companies around the world, such as Boeing Company, which has supported many research projects in aviation and manufacturing technology fields.

Table 6.18 displays some key statistical figures that illustrate the quality of research and innovation [81–84].

Figure 6.36 shows patent applications submitted by the University of Cambridge in different fields, while Fig. 6.37 shows the percentage of accepted patent applications.

Table 6.18 Some key statistical figures for the University of Cambridge

Number of Nobel laureates among professors and graduates	89
University budget	£1.7 billion
Endowments in 2017	£4.9 billion
Campus area	288 ha
Number of undergraduate students (2016)	12,340
Number of postgraduate students (2016)	7615
Teaching staff	1646
Number of administrative staff	3615

Fig. 6.36 Patents submitted by the University of Cambridge

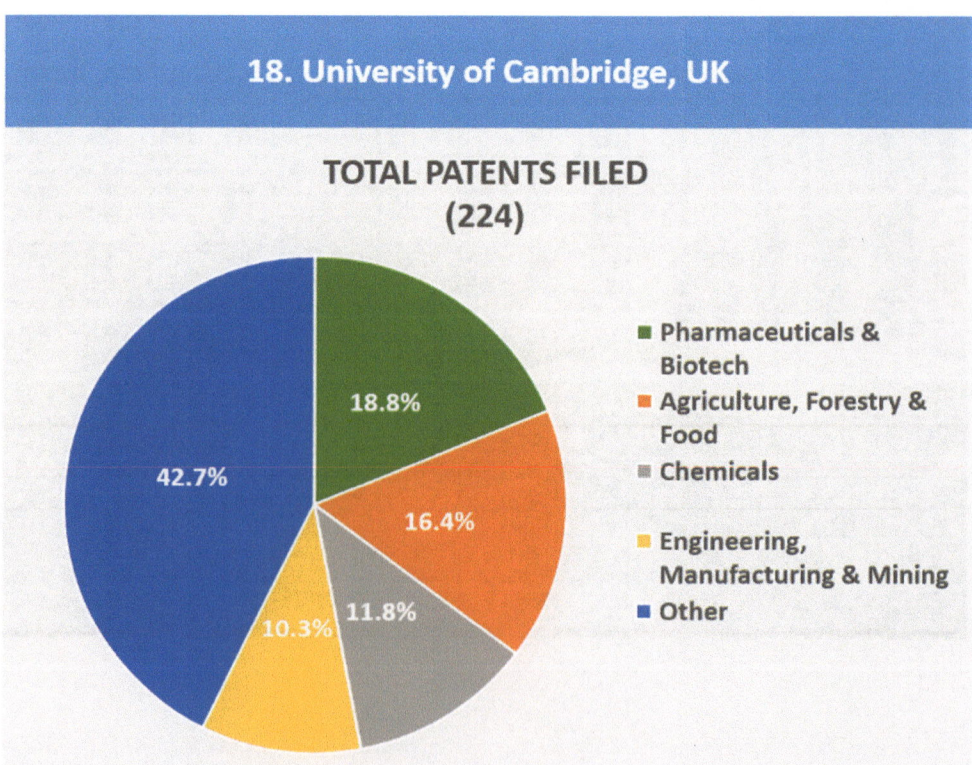

Fig. 6.37 Percentage of patents accepted for the University of Cambridge

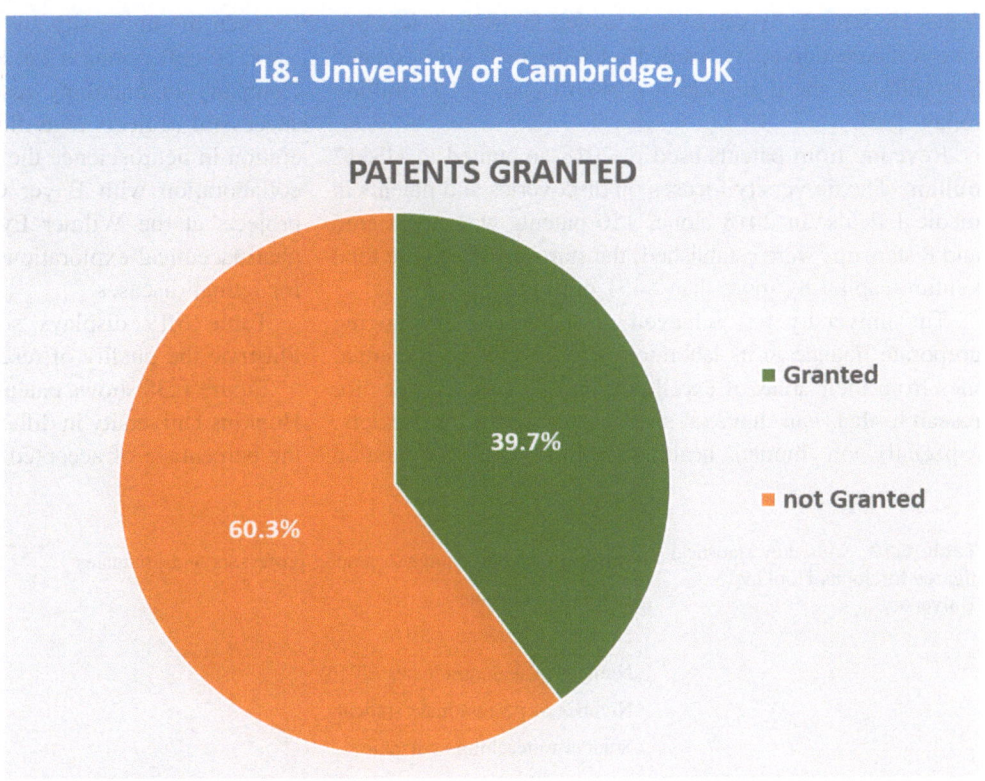

6.19 Johns Hopkins University—USA

Johns Hopkins University was founded in 1876; it is a private research university, named after the man who donated $7 million at the time (which is worth about $145 million today) [85].

Revenue from patents used in 2018 amounted to US$17 million. The university focuses on discoveries and patents in medical fields. In 2018 alone, 150 patents were registered and 8 start-ups were established; the start-ups raised the total venture capital by more than $451 million.

The university has achieved great success in bringing corporate finance to its laboratories to benefit these companies from their areas of excellence and provide avenues for research that can have a significant impact on society, especially on human health. Examples of cooperation between the university and giant companies include as follows: A collaboration agreement was signed with AbbVie Company in oncology research in 2016, resulting in six successful projects regarding tumors, with continued cooperation in neuroscience the following year. There is also the collaboration with Bayer Company in four joint research projects at the Wilmer Eye Institute, as well as work in pharmaceutical explorations at JHDD to develop treatments for retinal diseases.

Table 6.19 displays some key statistical figures that illustrate the quality of research and innovation [85–87].

Figure 6.38 shows patent applications submitted by Johns Hopkins University in different fields, while Fig. 6.39 shows the percentage of accepted patents.

Table 6.19 Some key statistical figures for Johns Hopkins University

Number of Nobel laureates among professors and graduates	27
Endowments value	Endowments value
Number of students (2016)	26,402
Number of undergraduates (2016)	5615
Number of postgraduate students	20,787
Student-to-teaching staff ratio	7:1

Fig. 6.38 Patents submitted by Johns Hopkins University

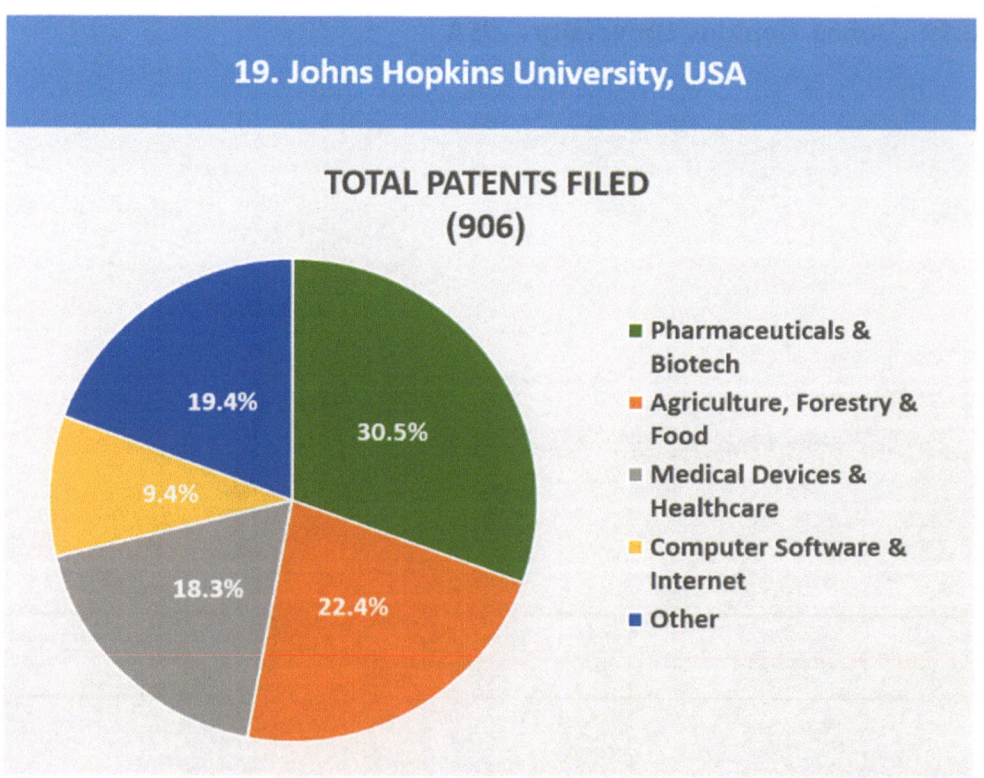

Fig. 6.39 Percentage of patents accepted for Johns Hopkins University

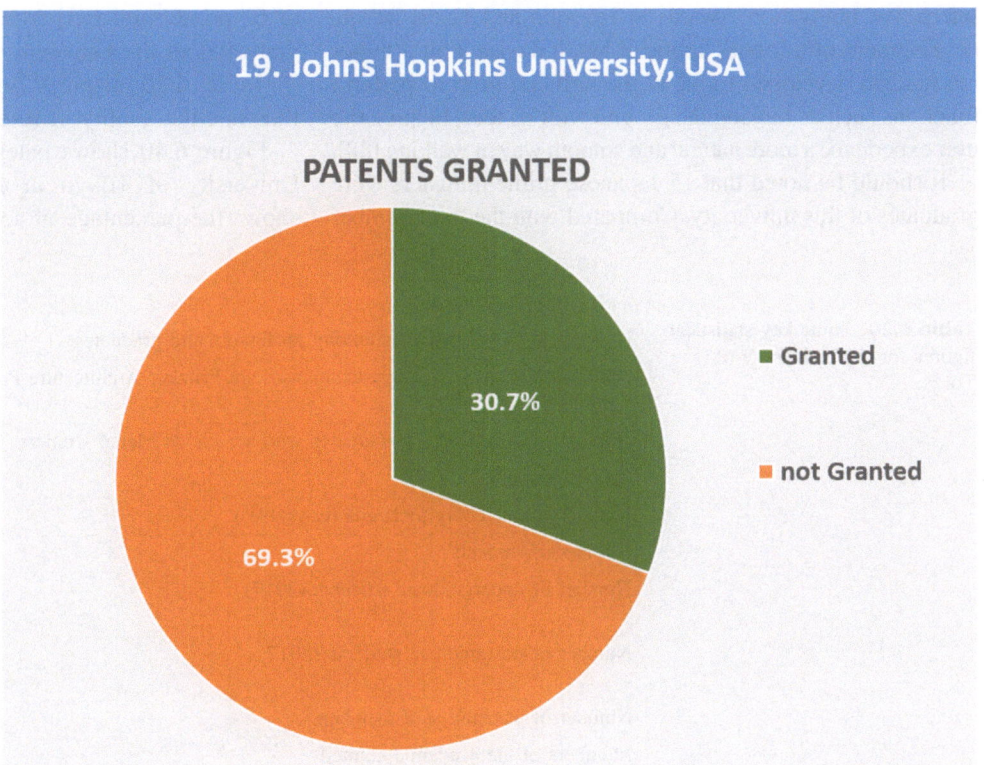

6.20 University of Tokyo—Japan

University of Tokyo was founded in 1877 as the first national university in Japan. It consists of three main campuses, as well as 14 university centers and 11 affiliated research institutes, including the Institute for Cosmic Ray Research and the Earthquake Research Institute; in addition to the University of Tokyo Institutes for Advanced Studies for the Physics and Mathematics of the Universe and the Integrated Research System for Sustainability Science.

Nearly 200 companies around the world collaborate with the University of Tokyo in research projects, including Johnson and Johnson, which helps in healthcare quality assessment, and Fujifilm Corporation, which supports the Laboratory of Drug Development.

Two graduates of the University of Tokyo won two different Nobel Prizes in 2015. In March 2017, students at the University of Tokyo won the Innovation for Students Award at the SXSW

Interactive Innovation Awards in Houston and Texas for the development of a robotic artificial leg, a device with a motor mechanism developed to allow the artificial limb to perform functions such as kicking the ground with its toes, helping the user experience a more natural and smooth way of walking [88].

It should be noted that 15 Japanese prime ministers were graduates of this university. Compared with the total number of 62 prime ministers, 24% of Japanese prime ministers are graduates of the University of Tokyo.

Table 6.20 displays some key statistical figures that illustrate the quality of research and innovation [88–90].

Figure 6.40 shows patent applications submitted by the University of Tokyo in different fields, while Fig. 6.41 shows the percentage of accepted patent applications.

Table 6.20 Some key statistical figures for the University of Tokyo

Number of Nobel laureates among professors and graduates	9
Number of professors or graduates who are Pritzker Architecture Prize winners	3
Number of professors or graduates who are Fields Medal winners	1
University budget (2017)	$2425 billion
Number of university locations (campuses)	3
Number of students	28,693
Number of undergraduate students (2017)	14,273 (49.7% of total students)
Number of postgraduate students (2017)	14,420 (50.3% of total students)
Number of teaching staff members	2609
Members of the academic council	6022
Number of administrative staff	5779

Fig. 6.40 Patents submitted by the University of Tokyo

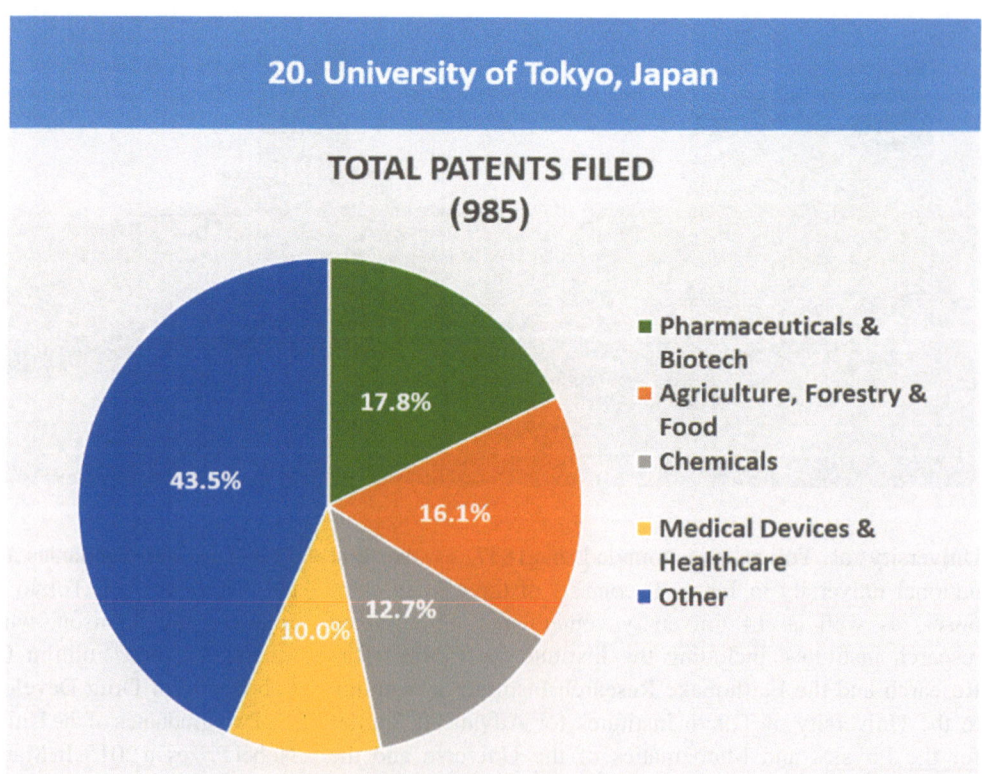

Fig. 6.41 Percentage of patents accepted for the University of Tokyo

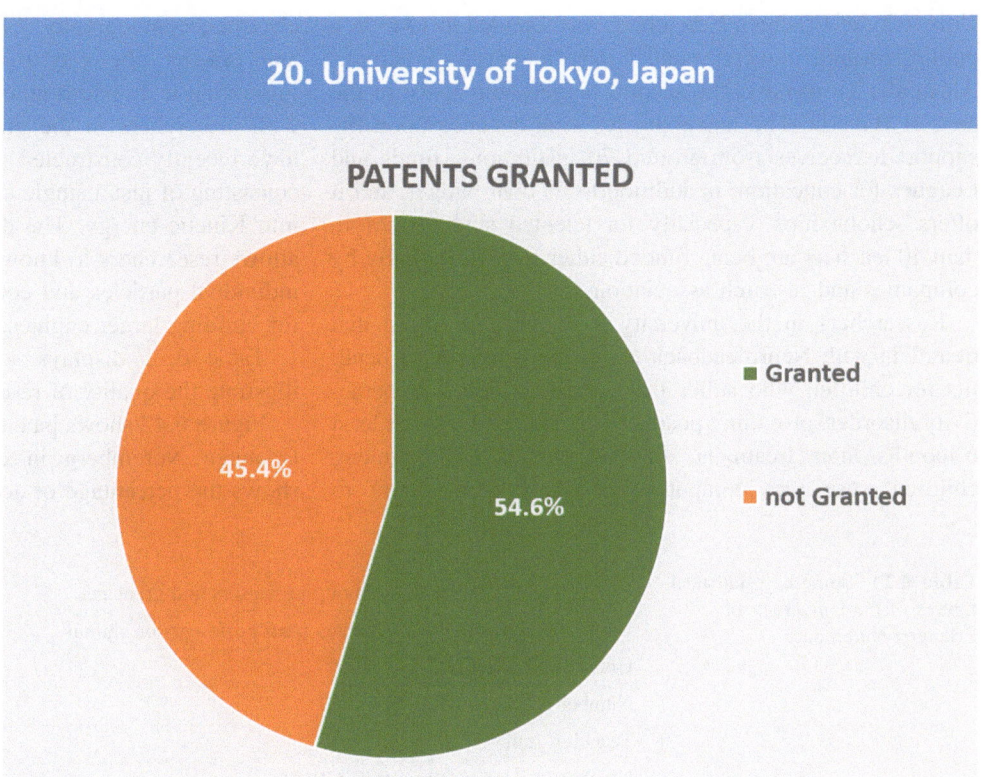

6.21   University of Erlangen Nuremberg—Germany

University of Erlangen Nuremberg was founded in 1743 as a public research university and is also known as Friedrich-Alexander University, the second biggest university in the area of Bavaria. The university research depends upon the support it receives from around 70 institutions, funds and Leagues for education, in addition to its own budget, and it offers scholarships, especially for talented students. More than 40 teachers are being funded either fully or partially by companies and research associations.

Researchers in the university have recently found that treatment with Neurofeedback can achieve long-term benefits for children who suffer from attention deficit hyperactivity disorder, providing positive effects that last for at least 6 months after treatment. In this method of treatment, children use a computer software connected to electroencephalography (EEG) to monitor the brain activity and practice the way of organizing their behavior and improving it. In addition, other researchers, in collaboration with researchers in the universities of Mainz and Kassel, have recently constructed the smallest engine in the world, consisting of just a single atom that can effectively turn heat into Kinetic energy. The development of such nanomotors allows researchers to know more about thermodynamics of individual particles and could help lead to new approaches for building larger engines.

Table 6.21 displays some key statistical figures that illustrate the quality of research and innovation [91–92].

Figure 6.42 shows patents submitted by the University of Erlangen Nuremberg in different fields, while Fig. 6.43 shows the percentage of accepted patents.

Table 6.21 Some key statistical figures of the University of Erlangen Nuremberg

Number of Nobel laureates among professors and graduates	4
Number of Leibniz prize laureates from professors or alumni	6
University budget	€543.1 million
Number of students (2015)	40,174
Teaching staff	4040
Number of administrative staff (2015)	2290

Fig. 6.42 Patents submitted by the University of Erlangen Nuremberg

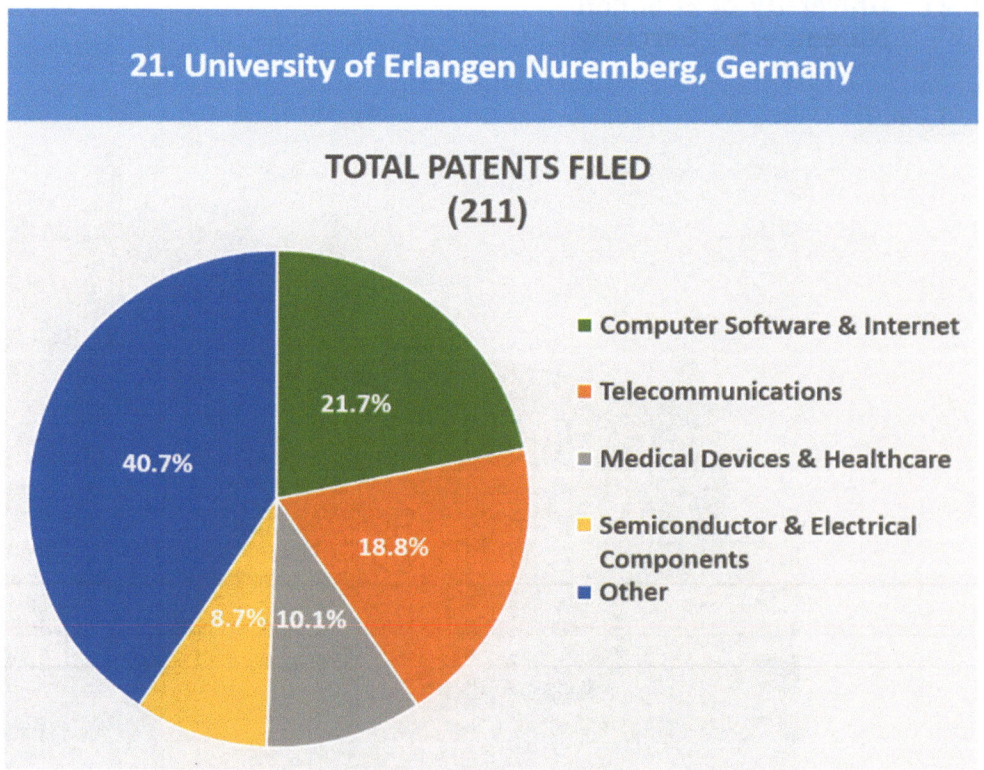

Fig. 6.43 Percentage of accepted patents of the University of Erlangen Nuremberg

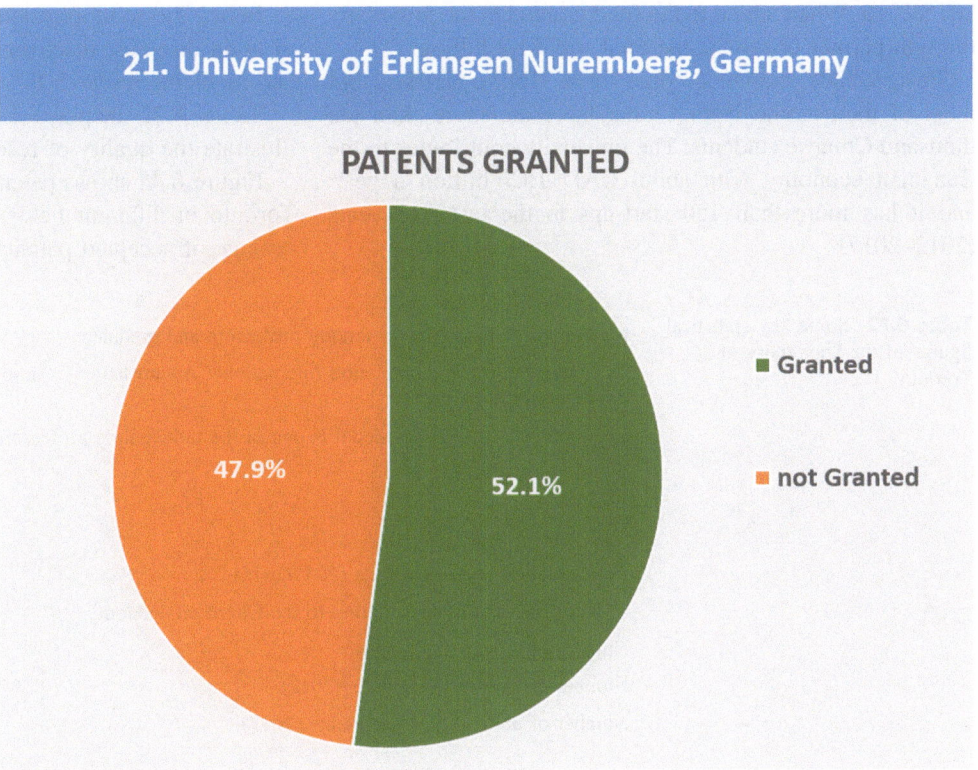

6.22 University of Toronto—Canada

The university was founded by the royal charter in 1827 as a royal university, and now it has three campuses in the area of Toronto.

In March 2017, University of Toronto launched an institute for artificial intelligence (Vector Institute of Artificial Intelligence), an independent non-profit research institute that conducts research and works on adopting these techniques and marketing them. The institute cooperates with companies, such as Google and Uber. University of Toronto is a founding member of MaRS Discovery District, a non-profit organization that helps entrepreneurs in both public and private sectors to grow their companies. Recently, the university has also allocated a new 14,000-ft^2 area in the campus to cooperate with entrepreneurs. The total value of research funds granted to the university's researchers was $391 million in the academic year

2015/2016. When added to the fund granted to the university through partner hospitals, the total was $1.2 billion.

Postgraduate students come from 168 countries, but most of them come from China; there are more than ten thousand Chinese students. The university contributes to the Canadian economy with about CAD 15.7 billion a year, and it has more than 150 start-ups in the last five years (2012–2017).

James Till, a biophysicist in the university and an expert in brain diseases, discovered stem cells in the early 1960s, and medicine research is still a main area of focus [93].

Table 6.22 displays some key statistical figures that illustrate the quality of research and innovation [93–96].

Figure 6.44 shows patents submitted by University of the Toronto in different fields, while Fig. 6.45 shows the percentage of accepted patents.

Table 6.22 Some key statistical figures of the University of Toronto

Number of Nobel laureates among professors and graduates	10
Number of Gairdner Foundation International Award laureates among professors or alumni	11
The establisher of Fields Medal is one of the university's professors	John Charles Fields
Endowments (2017)	CAD 2.8 billion ($2.22 billion)
University budget (2017)	$1 billion
Total value of research grants (2015/2016)	CAD 1.2 billion
University's annual contribution to the Canadian economy	CAD 15.7 billion
Number of university campuses	3
Number of university students (2017)	88,066
Number of undergraduate students (2017)	70,028 (80% of total students)
Number of postgraduate students (2017)	18,038 (20% of total students)
Number of teaching staff members	2547
Number of assistant professors	(14%)
Members of the academic council	5274
Number of administrative staff	4590

Fig. 6.44 Patents submitted by the University of Toronto

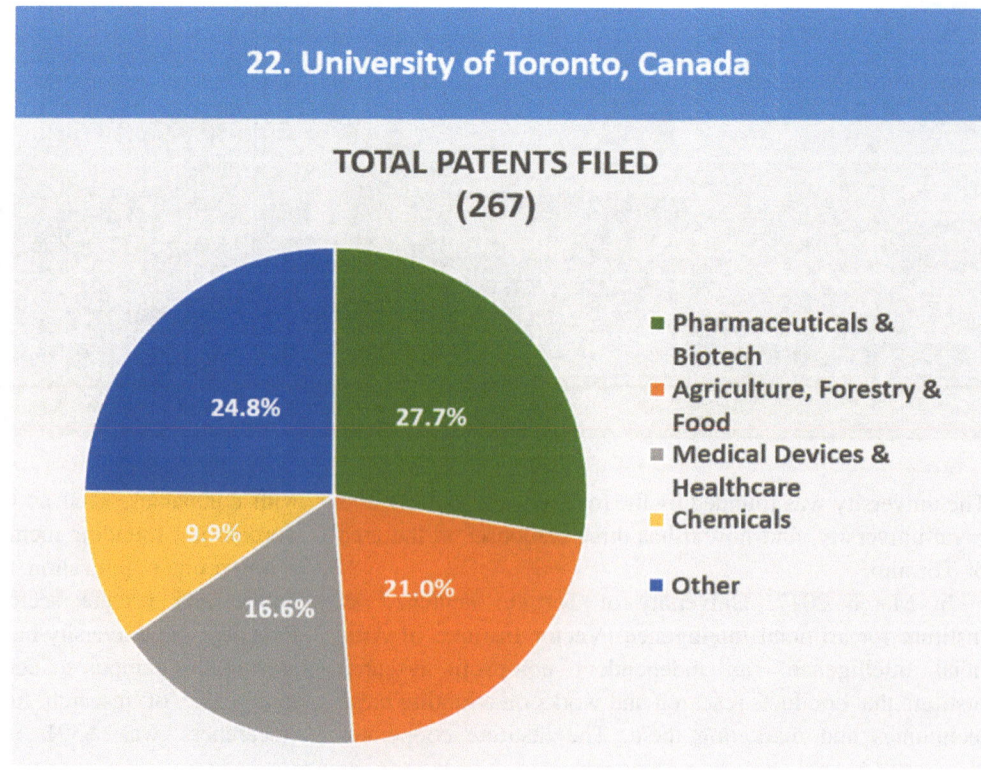

Fig. 6.45 Percentage of accepted patents of the University of Toronto

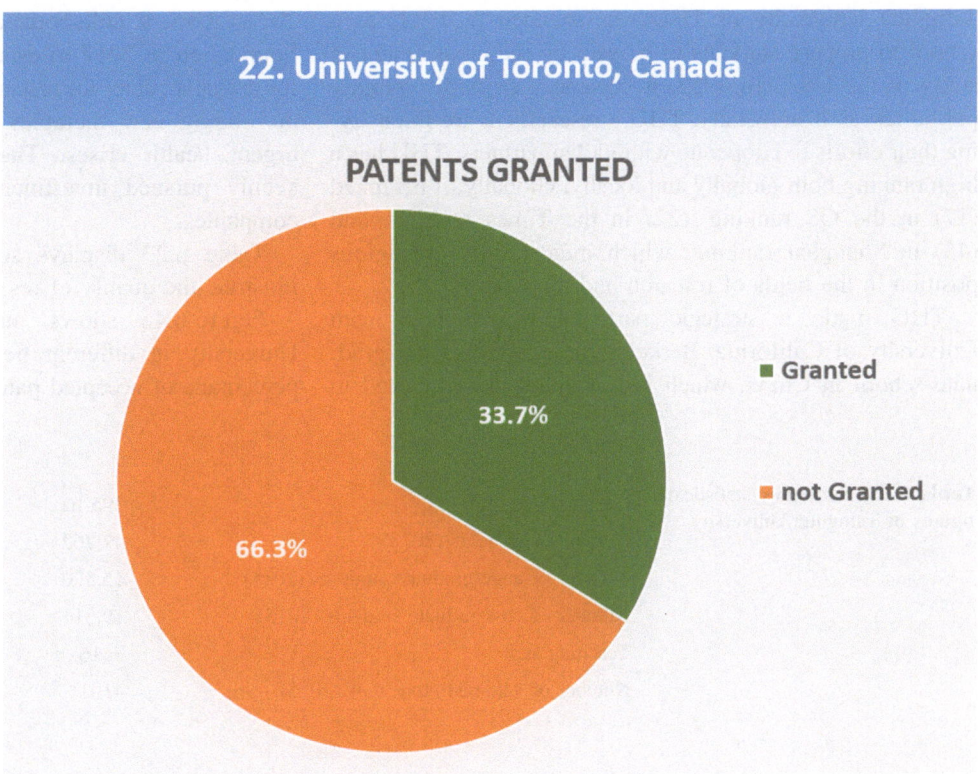

6.23 Tsinghua University—China

Tsinghua University or THU was founded in 1911, as a school to prepare students to be sent by the government to study in the USA. In 1952, it became a cross-disciplinary public research university. THU's researchers are intensifying their efforts to cooperate with global partners. THU has a high ranking both globally and locally. Globally, it is ranked (17) in the QS ranking, (22) in the Times ranking, and (45) in Shanghai ranking, which indicates its prestigious position in the fields of research and innovation.

THU made a strategic partnership agreement with University of California, Berkeley, to establish a postgraduate school in China, which is scheduled to be opened in 2021. Also, it collaborated with Bill and Melinda Gates Foundation in 2017 to establish the World Health Institute for drugs, a center for research and development that focuses on finding new therapies for some of the world's most urgent health crises. The university's investment arms keenly pursued investments in American semiconductor companies.

Table 6.23 displays some key statistical figures that illustrate the quality of research and innovation [97–98].

Figure 6.46 shows patents submitted by Tsinghua University in different fields, while Fig. 6.47 shows the percentage of accepted patents.

Table 6.23 Some key statistical figures of Tsinghua University

Campus area	395 ha
Number of students (2018)	47,762
Number of undergraduate students (2015)	15,570
Number of postgraduate students (2015)	19,311
Teaching staff	3416
Number of administrative staff (2015)	4101

Fig. 6.46 Patents submitted by Tsinghua University

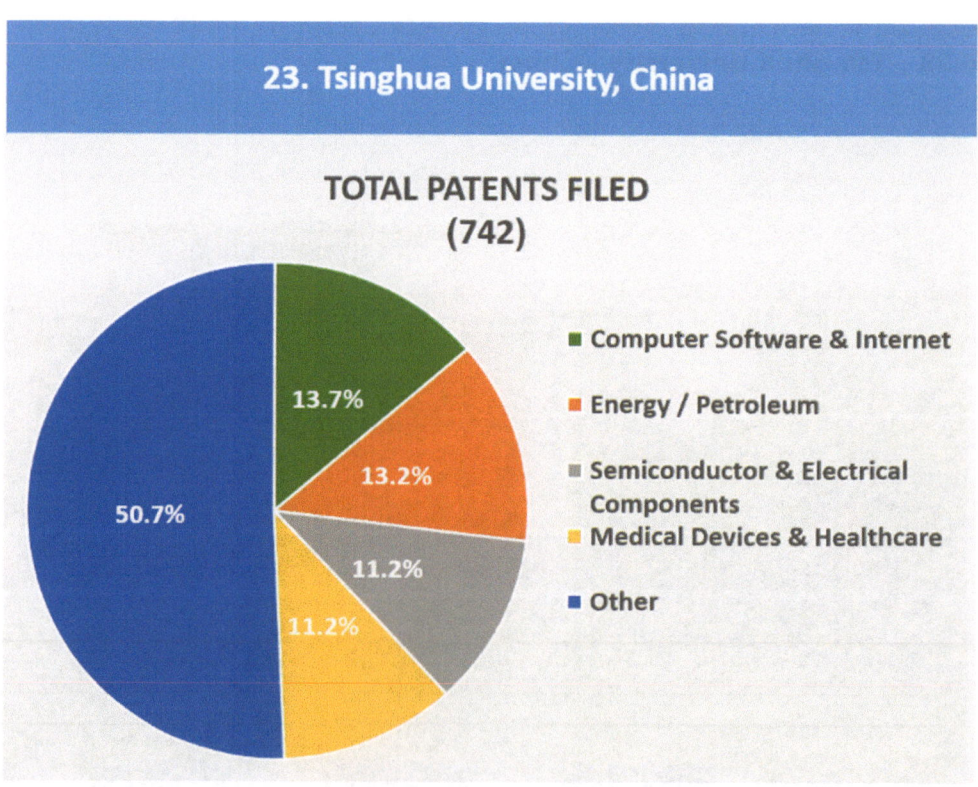

Fig. 6.47 Percentage of accepted patents of Tsinghua University

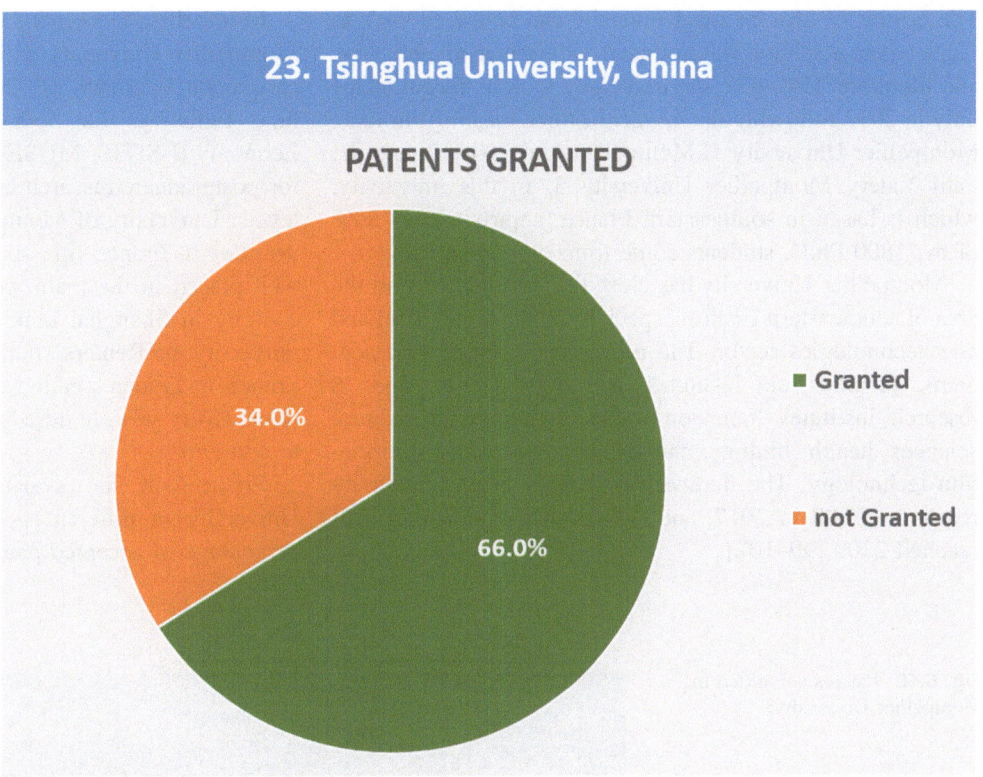

6.24   Montpellier University—France

The origins of Montpellier University can be traced back to 1289 when a school was founded to teach medicine, law, and literature. However, the university took its recent shape only in 2012 following the merger of three local institutions: Montpellier University 1, Montpellier University 2, and the Paul Valery Montpellier University 3. In this university, which is based in southeastern France, approximately 50% of the 1800 Ph.D. students come from other countries.

Montpellier University has close ties with industry in the area of southeastern France, especially in the biomedical and new technologies sector. The university includes nine academic faculties, six institutes, and two schools. The 76 research institutes focus on topics including agricultural sciences, health, biology, chemistry, physics, and information technology. The number of students in the university reached 47,000 in 2017, and the number of teaching staff reached 2300 [99–102].

Montpellier University has 586 patent families, and the Montpellier University of Excellence project (MUSE) was chosen on February 21, 2017, by an international jury to fund Initiatives for Science, Innovation, Territories and Economy (I-SITE). MUSE will provide a tremendous boost for postgraduate research on both national and international levels. University of Montpellier is one of the pioneer universities in France (the sixth French university). It is very well placed in the main world ranking, as it is the first in ecology in Shanghai ranking in 2018, the first innovative university in Reuters' ranking, and is ranked the fifth in France in Leiden ranking for quality of scientific publications. Every year, hundreds of international students choose to come to it.

Figure 6.48 shows patents submitted by Montpellier University in different fields, while Fig. 6.49 shows the percentage of accepted patents.

Fig. 6.48 Patents submitted by Montpellier University

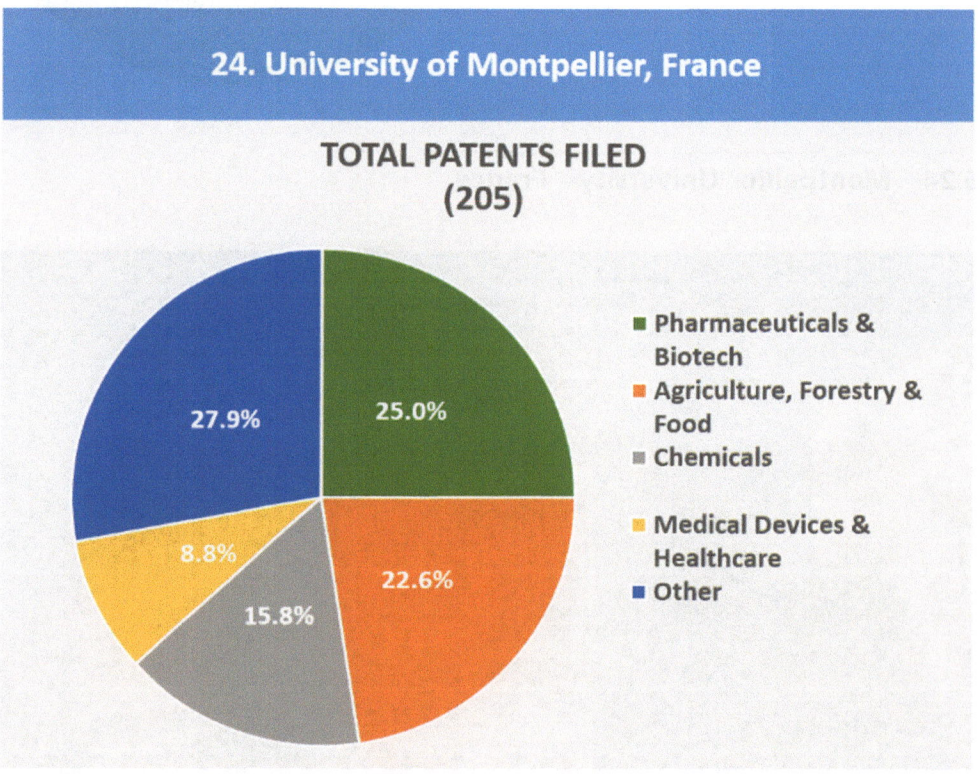

Fig. 6.49 Percentage of accepted patents of Montpellier University

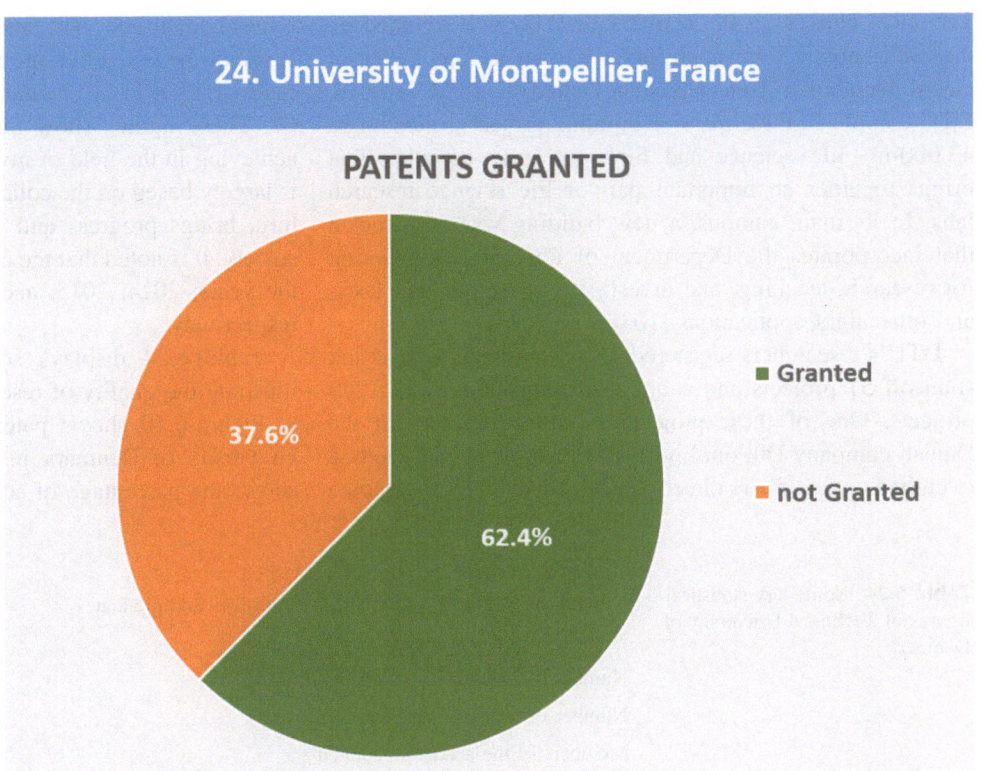

24. University of Montpellier, France

PATENTS GRANTED

37.6%

62.4%

- Granted
- not Granted

6.25 Technical University of Denmark—Denmark

Technical University of Denmark or DTU was founded by the well-known Danish physicist known as the father of electromagnetism (Hans Christian Ørsted) in 1829. In 2016, DTU completed three massive building projects, including a 43,000-m^2 life science and bioengineering complex that brings together an important part of life science research labs. In its main campus, a new building was constructed that incorporates the Department of Photonic Engineering for research, teaching, and investigating photons and laser, and other light applications [103].

DTU's researchers registered 152 inventions in 2014 and spun off 51 projects and cooperated with industry in 1249 projects. One of these projects is collaboration with the Danish company DuPont Lightstone, and involves a method to embed optical fibers directly into LED displays to become

robust and durable. The student-to-teaching staff ratio is 6: 1. It should be noted that the number of projects with industry reached 1324 in 2017, and the number of start-ups reached 67. These figures show that the success this university is achieving in the field of inventing and establishing start-ups is largely based on the collaboration with industry, which, in turn, brings progress and prosperity to the university and society. It is noted that the numbers of PhD degree holders in the years: 2014, 2015, and 2016 were 323, 358, and 359, respectively.

Table 6.24 displays some key statistical figures that illustrate the quality of research and innovation [103–105].

Figure 6.50 shows patents submitted by the Technical University of Denmark in different fields, while Fig. 6.51 shows the percentage of accepted patents.

Table 6.24 Some key statistical figures of Technical University of Denmark

Number of Nobel laureates among professors and graduates	2
Number of undergraduates (2016)	7197 (65% of total students)
Number of postgraduate students (2016)	3834 (35% of total students)
Number of teaching staff members	2003
Members of the academic council	1330
Number of administrative staff	1540
Number of projects carried out with industry (2015)	1249
Number of projects carried out with industry (2016)	1324
Number of start-ups (2016)	67

Fig. 6.50 Patents submitted by Technical University of Denmark

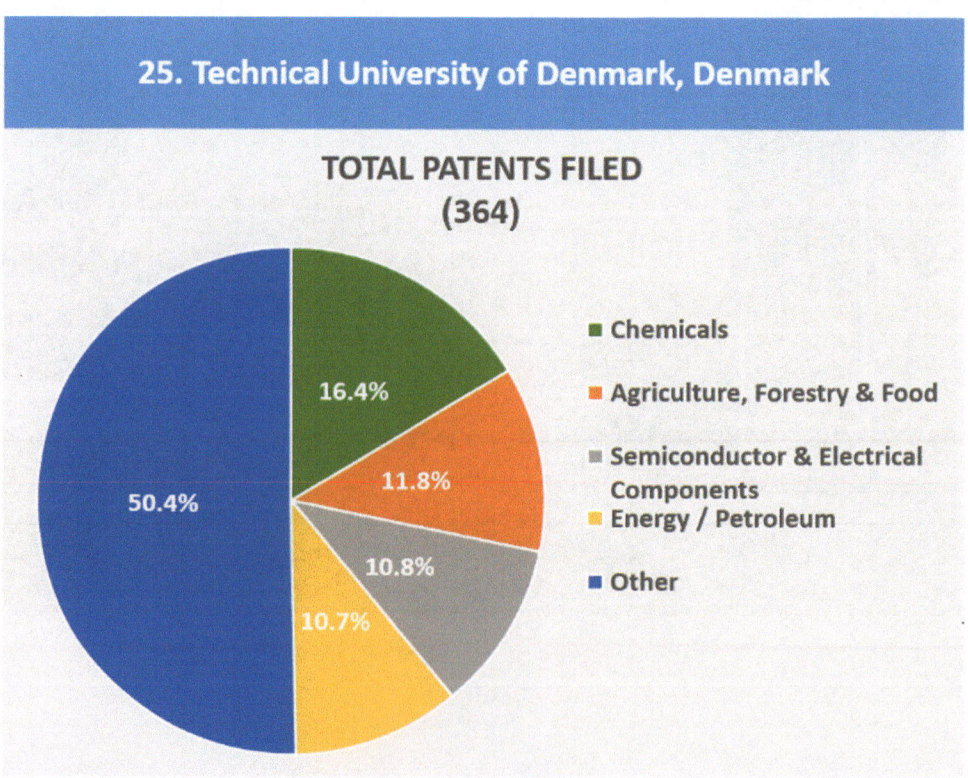

Fig. 6.51 Percentage of accepted patents of Technical University of Denmark

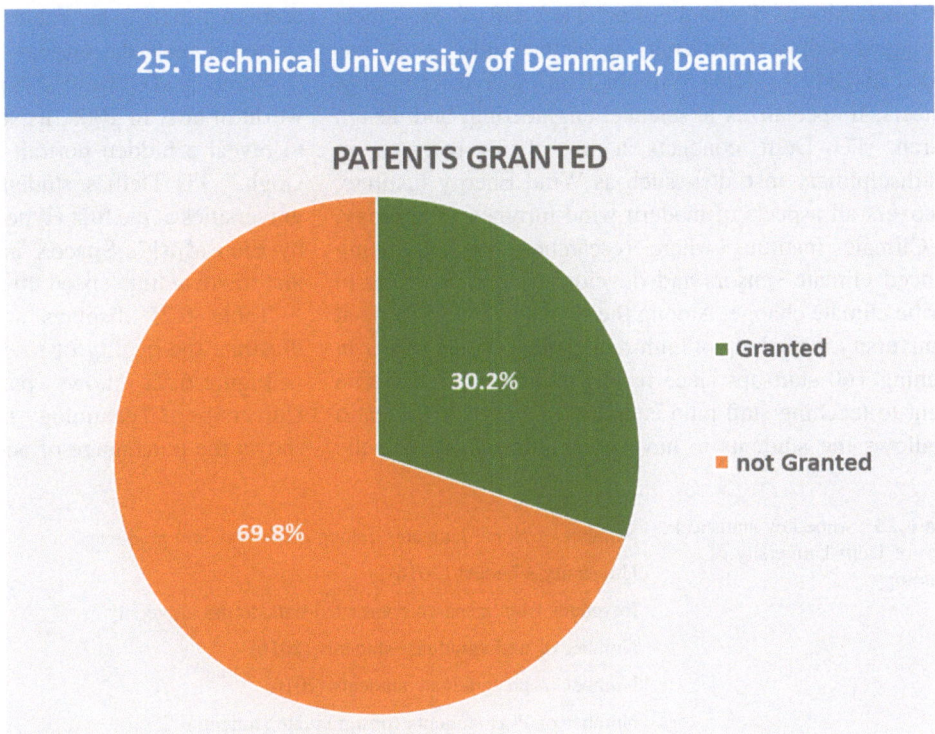

6.26   Delft University of Technology—Netherlands

Delft University of Technology or (TU Delft) is the oldest and biggest technical university in Netherlands, for it was founded in 1842 by King William II in order to train civil engineers. It specializes in science, engineering, and design research. TU Delft conducts a lot of its research in cross-disciplinary institutes, such as Wind Energy Institute, that covers all aspects of modern wind turbines technology, and Climate Institute, where researchers are producing advanced climate sensors and developing better models to describe climate change. Among the significant work done at the university is the help of high-tech entrepreneurs center in launching 160 start-ups since it was founded in 2005. The student-to-teaching staff ratio is approximately 4.4:1, a ratio that allows the students to have good interaction with the academic staff. The value of revenues from projects contracted outside the university is €184.6 million [106].

Some of TU Delft's expertise has also extended to the world of arts. In 2008, its scientists used X-ray spectroscope to reveal a hidden portrait under a painting for artist "Van Gogh." TU Delft's students have recently beat 26 other universities at the first Hyperloop Pod competition, organized by Elon Musk's SpaceX in Los Angeles, a test of vehicles that travel at high speed through a 1.2-km low-pressure tube.

Table 6.25 displays some key statistical figures that illustrate the quality of research and innovation [106–108].

Figure 6.52 shows patents submitted by the Delft University of Technology in different fields, while Fig. 6.53 shows the percentage of accepted patents.

Table 6.25 Some key statistical figures of Delft University of Technology

Number of Nobel laureates among professors and graduates	2
University's budget (2016)	€1246 billion ($1.47 billion)
Revenues from research contracted outside the university	€184.6 million
Number of undergraduate students (2016)	11,363 (54% of total students)
Number of postgraduate students (2016)	9860 (46% of total students)
Number of PhD students (postgraduate students)	1093
Number of teaching staff members	3375
Members of the academic council	2280
Number of administrative staff	2935
Number of start-ups from 2005 to 2017	160

Fig. 6.52 Patents submitted by Delft University of Technology

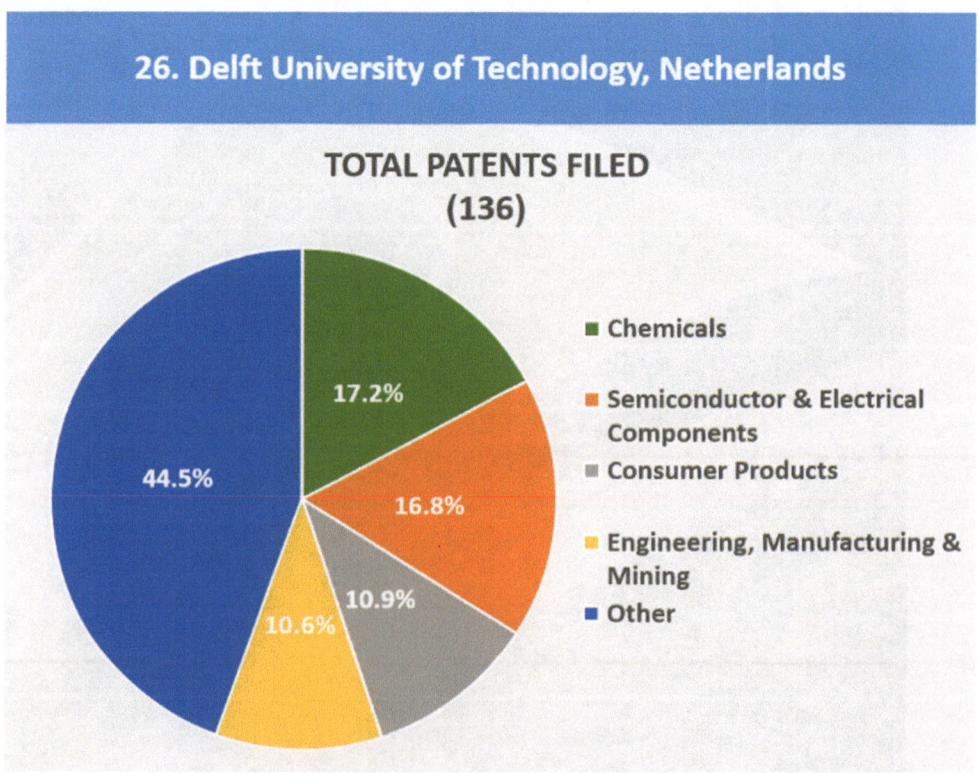

Fig. 6.53 Percentage of accepted patents of Delft University of Technology

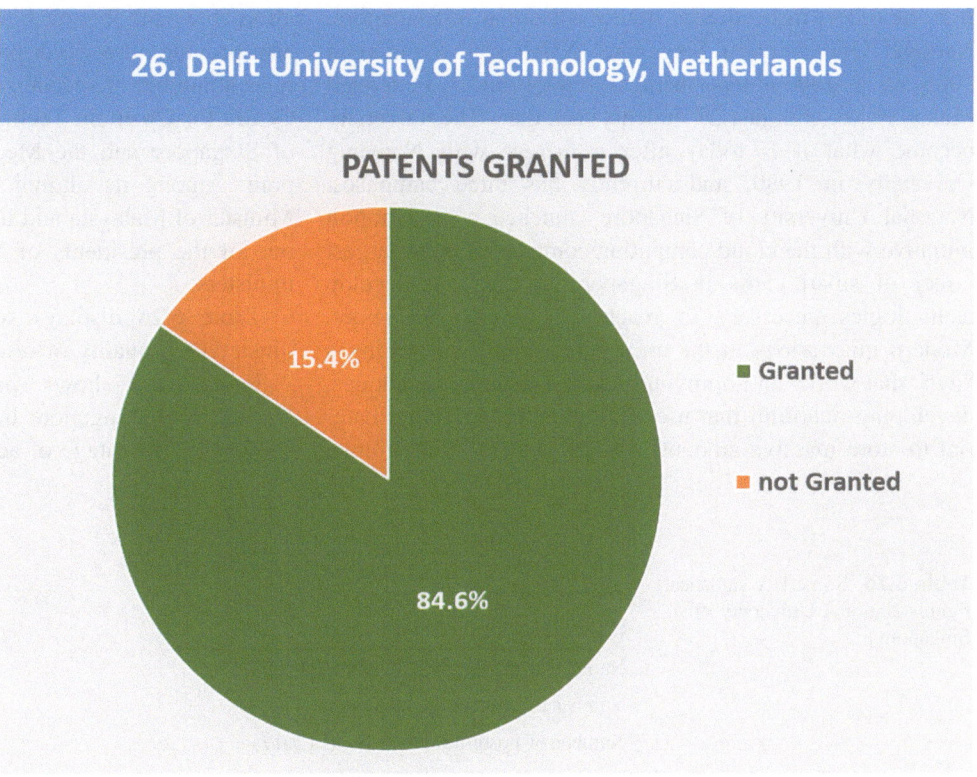

6.27   National University of Singapore—Singapore

It is the oldest institution of higher education in Singapore. National University of Singapore (NUS) was founded in 1905 as a government medical school under Federated Malay States, with just 23 students back then. The university became what it is today after a merger with Nanyang University in 1980, and currently has three campuses. National University of Singapore launched a cooperation initiative with the cloud computing company and the largest issuer of smart cards in Singapore in order to develop technologies necessary to reach a Cashless Singapore. Modern innovations in the university include innovating a "dye" that works on improving medical imaging techniques, developing nanofilm that uses tiny moving magnetic material to store massive amounts of data; and creating simple, affordable nanofiber air filters that not only clean the air, but also provide ultra violet protection. The university incorporates a number of specialized research centers, including the Centre for Quantum Technologies, Cancer Science Institute of Singapore and the Mechanobiology Institute of Singapore. Among its alumni are Mahathir Mohamad, Prime Minister of Malaysia and the pioneer of its renaissance, also one of the presidents of Singapore and two of its prime ministers.

Table 6.26 displays some key statistical figures that illustrate the quality of research and innovation [109–111].

Figure 6.54 shows patents submitted by National University of Singapore in different fields, while Fig. 6.55 shows the percentage of accepted patents.

Table 6.26 Some key statistical figures National University of Singapore

Endowments (2017)	S$3.73 billion ($2.76 million)
Campuses	3
Number of university's students (2017)	39,536
Number of undergraduate students (2017)	29,130 (74% of total students)
Number of postgraduate students (2017)	10,406 (26% of total students)
Number of teaching staff members	2196
Members of the academic council	2820

Fig. 6.54 Patents submitted by National University of Singapore

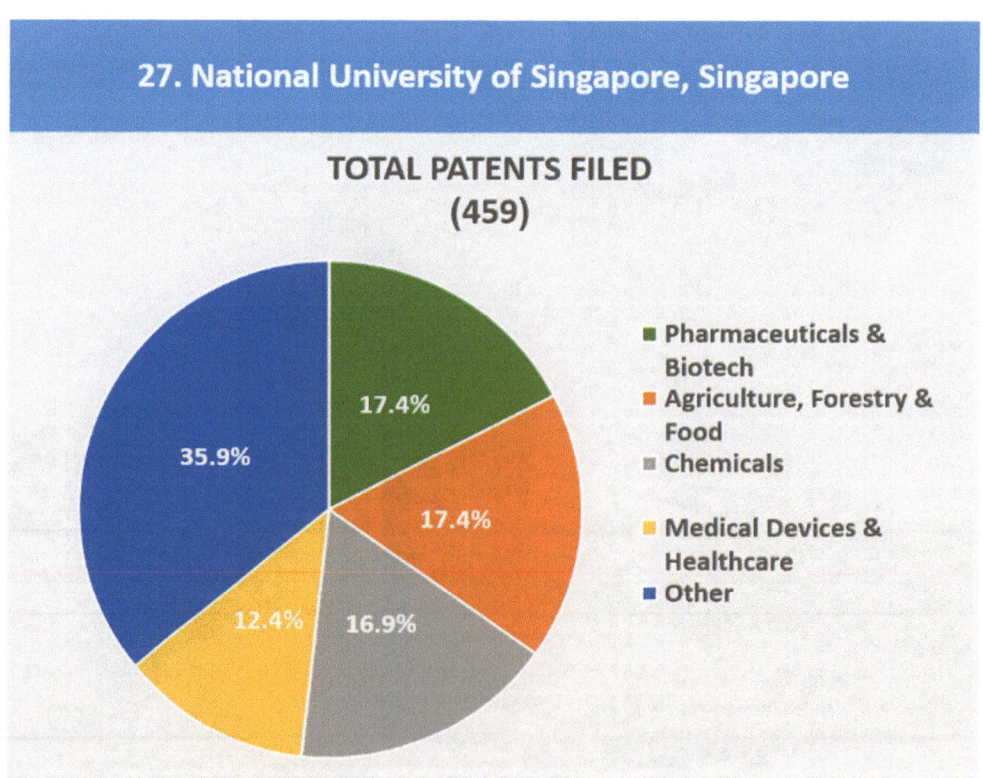

Fig. 6.55 Percentage of accepted patents of National University of Singapore

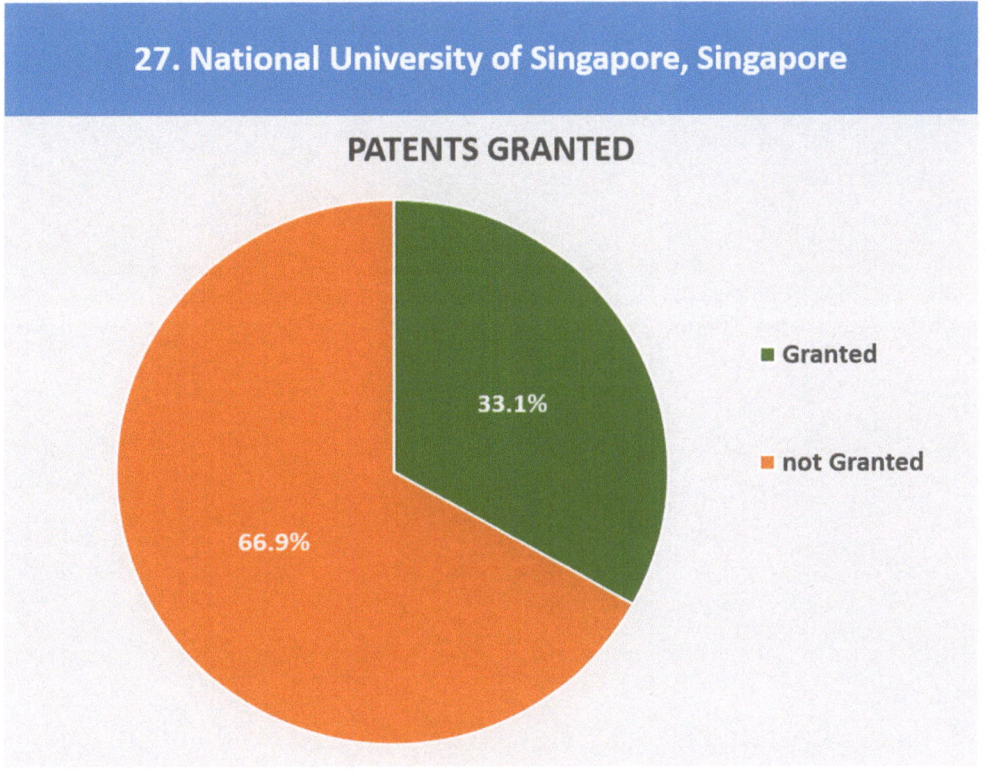

Statistics of Patents of the World's Most Innovative Universities

7.1 World Universities

Patents are the significant factor in ranking universities in terms of innovation. Therefore, in this chapter, we are presenting statistics on patents of different ranked universities in order to encourage other universities to enter the ranking if they work on registering patents. We are especially drawing the attention to the percentage of the number of registered patents in comparison with the number of patents submitted by universities, as we notice that this percentage is lower than 50% for a lot of American universities while it is higher for Asian universities. Table 7.1 shows the number of patents submitted to patent offices by world universities, the number of accepted patents, and the percentage between the number of registered patents and the number of submitted patents in 2018 [27]. Note that the table is based on world university rankings order shown in Table 7.1, and also on all registered patents of any university in a globally-recognized patent office.

The table shows that a limited number of universities succeeded in getting 80% of the patents they submitted to be accepted while the majority of universities did not exceed 50%, and some universities only got the quarter out of what they submitted to the patent offices in the USA, UK, and Japan to be accepted.

7.2 European Universities

Table 7.2 shows the number of patents submitted to patent offices by European universities, the number of accepted patents, and the percentage of the number of registered patents to the number of submitted patents in 2018 [28].

7.3 Asia-Pacific Universities

Table 7.3 shows the number of patents submitted to patent offices by Asia-Pacific universities, the number of accepted patents, and the percentage of the number of registered patents to submitted patents in 2018 [29].

© The Author(s) 2021
The Leading Worldâs Most Innovative Universities,
https://doi.org/10.1007/978-3-030-59694-1_7

Table 7.1 Number of patents filed by the institutions, the patents granted and the ratio for the World's Most Innovative Universities (2018)

Rank	Institution	Patents filed	Patents granted	Ratio (%)
1	Stanford University	691	287	41.50
2	Massachusetts Institute of Technology (MIT)	1480	666	45.00
3	Harvard University	977	296	30.30
4	University of Pennsylvania	557	174	31.20
5	University of Washington	536	179	33.40
6	University of Texas System	1023	352	34.40
7	KU Leuven	300	120	40.00
8	Imperial College London	297	99	33.30
9	University of North Carolina Chapel Hill	368	131	35.60
10	Vanderbilt University	231	106	45.90
11	Korea Advanced Institute of Science & Technology (KAIST)	1000	798	79.80
12	Ecole Polytechnique Federale de Lausanne	213	87	40.80
13	Pohang University of Science & Technology (POSTECH)	372	293	78.80
14	University of California System	2533	889	35.10
15	University of Southern California	294	78	26.40
16	Cornell University	534	228	42.70
17	Duke University	344	114	33.10
18	University of Cambridge	224	89	39.70
19	Johns Hopkins University	906	278	30.70
20	University of Tokyo	985	538	54.60
21	California Institute of Technology	596	350	58.70
22	Osaka University	583	307	52.70
23	University of Michigan System	608	268	44.10
24	Northwestern University	322	143	44.40
25	University of Wisconsin System	266	181	68.00
26	Kyoto University	688	360	52.30
27	University of Minnesota System	259	107	41.30
28	University of Illinois System	349	160	45.80
29	Georgia Institute of Technology	284	128	45.10
30	University of Utah	308	117	38.00
31	University of Erlangen Nuremberg	211	110	52.10
32	Ohio State University	208	106	51.00
33	Columbia University	632	169	26.70
34	Seoul National University	869	650	74.80
35	University of Toronto	267	90	33.70
36	Tohoku University	625	382	61.10
37	University of Pittsburgh	294	120	40.80
38	Yale University	273	89	32.60
39	Sungkyunkwan University	241	186	77.20
40	University of Oxford	469	123	26.20
41	University of Colorado System	318	108	34.00

(continued)

Table 7.1 (continued)

Rank	Institution	Patents filed	Patents granted	Ratio (%)
42	Tufts University	173	45	26.00
43	Baylor College of Medicine	229	52	22.70
44	Tsinghua University	742	490	66.00
45	Technical University of Munich	177	79	44.60
46	Kyushu University	554	286	51.60
47	Tokyo Institute of Technology	348	210	60.30
48	University College London	232	69	29.70
49	ETH Zurich	275	81	29.50
50	Purdue University System	291	119	40.90
51	University of Chicago	186	73	39.20
52	Oregon Health & Science University	153	54	35.30
53	University of Manchester	123	37	30.10
54	Indiana University System	235	87	37.00
55	Universite de Montpellier	205	128	62.40
56	University of Munich	107	45	42.10
57	Technical University of Denmark	364	110	30.20
58	Emory University	227	72	31.70
59	Peking University	437	258	59.00
60	Sorbonne University	329	164	49.80
61	University of British Columbia	199	69	34.70
62	Delft University of Technology	136	115	84.60
63	National University of Singapore	459	152	33.10
64	Princeton University	164	74	45.10
65	University of Zurich	167	61	36.50
66	Hanyang University	431	341	79.10
67	Case Western Reserve University	189	69	36.50
68	Yonsei University	539	407	75.50
69	Rutgers State University New Brunswick	229	989	432
70	Boston University	137	43	31.40
71	University of Massachusetts System	237	89	37.60
72	Johannes Gutenberg University of Mainz	88	34	38.60
73	Wake Forest University	130	52	40.00
74	Keio University	184	102	55.40
75	Korea University	508	435	85.60

Table 7.2 Number of patents filed by the institutions, the patents granted, and the ratio for European Most Innovative Universities (2018)

Rank	Institution	Patents filed	Patents granted	Ratio (%)
1	KU Leuven	300	120	40.00
2	Imperial College London	297	99	33.30
3	University of Cambridge	224	89	39.70
4	Federal Institute of Technology in Lausanne (EPFL)	213	87	40.80
5	University of Erlangen Nuremberg	211	110	52.10
6	Technical University of Munich	177	79	44.60
7	University of Manchester	123	37	30.10
8	University of Munich	107	45	42.10
9	Technical University of Denmark	364	110	30.20
10	Swiss Federal Institute of Technology Zurich	275	81	29.50
11	University College London	232	69	29.70
12	Delft University of Technology	136	115	84.60
13	University of Zurich	167	61	36.50
14	University of Oxford	469	123	26.20
15	University of Basel	57	22	38.60
16	University of Montpellier	205	128	62.40
17	Leiden University	75	38	50.40
18	Pierre & Marie Curie University—Paris 6	329	164	49.80
19	University of Paris Descartes—Paris 5	166	58	34.90
20	Ruprecht Karl University Heidelberg	144	47	32.60
21	Johannes Gutenberg University of Mainz	88	34	38.60
22	Free University of Berlin	115	37	32.20
23	Eindhoven University of Technology	50	25	50.00
24	University of Freiburg	167	81	48.50
25	University of Paris Sud—Paris 11	163	86	52.80
26	Charité Medical University of Berlin	85	23	27.10
27	Humboldt University of Berlin	111	36	32.40
28	Grenoble Alpes University	146	71	48.50
29	Dresden University of Technology	209	122	58.40
30	University of Bordeaux	169	95	56.20
31	Karlsruhe Institute of Technology	183	95	51.90
32	University of Oslo	92	27	29.30
33	Ghent University	235	107	45.50
34	University of Birmingham	93	30	32.30
35	University of Claude Bernard—Lyon 1	319	211	66.10
36	University of Glasgow	59	18	30.50
37	Queen Mary University London	71	26	36.60
38	King's College London	93	23	24.70
39	Technical University of Berlin	111	59	53.20
40	RWTH Aachen University	157	58	36.90
41	University of Strasbourg	166	63	38.00
42	Free University of Brussels	97	31	32.00
43	University of Copenhagen	95	23	24.20

(continued)

Table 7.2 (continued)

Rank	Institution	Patents filed	Patents granted	Ratio (%)
44	Polytechnic University of Milan	107	81	75.70
45	University of Edinburgh	111	28	25.20
46	Grenoble Institute of Technology	75	60	80.00
47	Vrije University of Brussels	72	27	37.50
48	Utrecht University	77	65	83.80
49	Ecole Polytechnique	91	75	82.40
50	Goethe University Frankfurt	57	17	29.80
51	University of Paris Diderot—Paris 7	164	52	31.70
52	University of Munster	99	34	34.30
53	Cardiff University	62	29	46.80
54	Catholic University of Louvain	81	21	25.90
55	Hannover Medical School	51	26	51.00
56	Erasmus University Rotterdam	89	19	21.70
57	University of Amsterdam	72	27	37.50
58	University of Dundee	61	16	26.20
59	University of Aix-Marseille	241	106	44.00
60	University of Leicester	56	20	35.70
61	Saarland University	58	29	50.00
62	University of Sheffield	69	14	20.30
63	Eberhard Karls University of Tubingen	93	34	36.60
64	Vienna University of Technology	104	76	73.10
65	Trinity College Dublin	83	18	21.70
66	University of Milan	50	33	66.00
67	University of Paul Sabatier—Toulouse III	145	71	49.00
68	University of Leeds	64	22	34.40
69	University of Barcelona	93	57	61.30
70	University of Southampton	79	30	38.00
71	University of Stuttgart	80	33	41.30
72	University of Wurzburg	51	19	37.30
73	University of Lorraine	104	83	79.80
74	University of Geneva	55	23	41.80
75	University of Twente	63	31	49.20

Table 7.3 Number of patents filed by the institutions, the patents granted, and the ratio for Asia-Pacific most innovative universities (2018)

Rank	Institution	Patents filed	Patents granted	Ratio
1	Korea Advanced Institute of Science and Technology (KAIST)	1000	798	79.80
2	University of Tokyo	985	538	54.60
3	Pohang University of Science and Technology (POSTECH)	372	293	78.80
4	Seoul National University	869	650	74.80
5	Tsinghua University	742	490	66.00
6	Osaka University	583	307	52.70
7	Kyoto University	688	360	52.30
8	Sungkyunkwan University	241	186	77.20
9	Tohoku University	625	382	61.10
10	National University of Singapore	459	152	33.10
11	Hanyang University	431	341	79.10
12	Peking University	437	258	59.00
13	Yonsei University	539	407	75.50
14	Kyushu University	554	286	51.60
15	Korea University	508	435	85.60
16	Tokyo Institute of Technology	348	210	60.30
17	Fudan University	137	77	56.20
18	Keio University	184	157	85.40
19	Shanghai Jiao Tong University	187	97	51.90
20	Gwangju Institute of Science and Technology	174	144	82.80
21	Zhejiang University	213	132	62.00
22	Chinese University of Hong Kong	74	46	62.20
23	Hokkaido University	291	169	58.10
24	Kyung Hee University	246	197	80.10
25	Monash University	165	34	20.60
26	Nanyang Technological University	423	168	39.70
27	Ajou University	183	153	83.60
28	Huazhong University of Science and Technology	174	132	75.90
29	Hiroshima University	165	110	66.70
30	Kumamoto University	123	73	59.30
31	Nagoya University	391	201	51.40
32	Beijing University of Chemical Technology	80	63	78.80
33	East China University of Science and Technology	129	74	57.40
34	Tokyo Medical & Dental University (TMDU)	121	37	30.60
35	Tianjin University	167	103	61.70
36	University of Sydney	162	44	27.20
37	Ewha Woman's University	171	135	78.90
38	Hong Kong University of Science and Technology	111	49	44.10
39	University of Auckland	114	41	36.00
40	Shinshu University	124	74	59.70
41	South China University of Technology	253	137	54.20
42	University of Queensland	161	49	30.40

(continued)

Table 7.3 (continued)

Rank	Institution	Patents filed	Patents granted	Ratio
43	Kanazawa University	67	42	62.70
44	China University of Petroleum	75	53	70.40
45	University of Melbourne	89	26	29.20
46	Southeast University China	188	127	67.60
47	University of Hong Kong	122	71	58.20
48	University of Tsukuba	108	56	51.90
49	Catholic University of Korea	166	132	79.50
50	Nanjing University	120	84	70.00
51	University of New South Wales Sydney	118	47	39.80
52	Chiba University	95	58	61.10
53	Xi'an Jiaotong University	113	81	71.70
54	Chonbuk National University	168	136	81.00
55	Chonnam National University	117	105	89.70
56	China University of Mining and Technology	315	236	74.90
57	Pusan National University	164	123	75.00
58	Dalian University of Technology	97	68	70.10
59	kayama University	151	97	64.20
60	Kyungpook National University	245	196	80.00
61	Harbin Institute of Technology	81	63	77.80
62	Nankai University	60	39	65.00
63	Chung-Ang University	101	95	94.10
64	Inha University	58	50	86.20
65	Sun Yat-sen University	107	47	43.90
66	Sichuan University	73	42	57.50
67	Shandong University	71	51	71.80
68	University of Electronic Science and Technology of China	57	36	63.20
69	University of Ulsan	93	66	71.00
70	Waseda University	102	52	51.00
71	Indian Institutes of Technology System (IIT)	192	90	46.90
72	Kobe University	96	48	50.00
73	Konkuk University	86	67	77.90
74	Xiamen University	66	35	53.00
75	Tongji University	84	40	47.60

Arab universities, in order to be among the world's innovative universities, all the personnel thereof should be informed and educated about the following factors:

8.1 Scientific Research Importance

It is significant to be aware of the scientific research importance as a source of social and economic development. The term of innovative or the most innovative as a description of the university's future is the main focus of this book, so that the ambitious and talented youth feel open to new knowledge, science, economy, creativity, and arts sustainably; enabling the university to be a pioneer in our Arab and Islamic world. The time is ripe for the universities to be regarded as productive rather than consuming facilities because they present human resources, the greatest wealth of any countries. Therefore, the universities are required to provide researches and innovations that would lead to intellectual, economic, and social leadership. To achieve this target, our universities shall pioneer to embrace creative students. They have to draw students and innovative teaching staff who will come up with creative and innovative production through which the economic, social, and health affairs of community may flourish and develop. As known, the world's prestigious universities are the key catalyst and producer of innovation, business, and projects through generating new ideas. Accordingly, they contribute to supporting the economic growth business. This, of course, is not limited to the economic benefit of their research activities, but also includes, most importantly, the number of graduates as being gifted individuals contributing to the current technological and economic development.

To achieve these goals, It has become the duty of universities to train all teaching staff to recognize, learn to protect, and benefit from intellectual property when necessary, and to provide investors capable of supporting them financially. Such an academic and professional effort not only generates intellectual property but also converts it to products of commercial value that can be licensed to establish start-ups, generating new products and businesses of long-term commercial value. It is true that the complexities and risks of achieving this level of success cannot be ignored, but the reward of proper performance can also be significant. Hence, the guidance and advice shall be sought from those who have succeeded in dealing with this field, and who have direct experience in the establishment of high-tech companies of their own intellectual property.

The support, given by prestigious universities to their students and teaching staff to enter the business field on a personal level, became notably trending and is receiving a growing interest, especially in terms of education and research. Furthermore, the university, whenever having available capabilities, shall offer its facilities to enable businessmen among students and teaching staff to conduct their distinctive researches and refine their ideas. Also, the facilities and expertise should be available for outside businesses, enabling positively benefiting from ideas of research and development [112].

One of the criteria of evaluating universities on a global scale is the scientific research and the universities' innovations. Excellence in this field is a real criterion of the progress of countries and the advancement of their societies. Therefore, the scientific research conducted by universities has become a necessity to reach both of creativity and excellence for achieving the sustainable development due to how scientific research may solve many economic, health, educational, and social problems in accordance with correct scientific bases. In light of that, it was only natural for universities to pay great attention and direct their activities to train their professors and students to master the scientific research methods to acquire skills enabling them to add new knowledge to the balance of human intellect.

The acceleration of technological developments in this time poses a challenge to Arab countries, whether poor or high-income countries, as they have gained this rise in

The Leading Worldâ s Most Innovative Universities,
https://doi.org/10.1007/978-3-030-59694-1_8

income only because of their natural resources rather than the possession or generation of advanced technologies. This challenge is the necessity of working hard to promote the research, development, and innovation. These countries might face what hinders the full use of their potential power in their way toward excellence in innovation. Thus, they have to overcome these obstacles to live up to the challenge represented in turning their economy into a knowledge and innovation-based economy by means of developing and stimulating both the physical and human aspects of innovation system [113–115].

To be innovative, the culture of innovation must be prevailed in the university. As known, the peoples' cultures differ from one country to another; the Chinese culture is different from the Japanese one, and cultures may differ from one institution to another or from a university to another in one country, although the culture elements are similar. These differences do not mean that one culture is better than the other.

But why are some universities innovative and creative while others are not? Why would some institutions be perceived as creative while others as not creative? We do not believe that the answer depends only on the organizational structure of university or institution, nor on the financial resources. For the culture, adapted by the institution's staff is one of the aspects that make an institution effective in the fields of innovation and creativity. The culture has become an integral part of researches in public and private institutions. It is related to the criteria, values, and principles that are created, formed, and sustained in the institution. Although organizational culture is a broad and unspecified term, the researchers agree that a certain number of ideas may be examined when one wants to understand the culture of an institution. The following points determine the framework upon which the institution's culture recognition depends [115]:

1. Mission,
2. Environment,
3. Leadership,
4. Strategy,
5. Information,
6. Socialization.

The institution's mission refers to how all staff therein identify its overarching principles, as the institution's mission provides the meaning, guidance, and goal. Moreover, the institution's mission is usually determined by its history and surrounding environment; and the concept of the environment is variable and must be considered, redefined, and interpreted constantly.

Leadership is also a cultural construction as leaders set mechanisms enabling them to work and somehow communicate in the institution, through interaction and communication methods with each other. These mechanisms also allow them to know how decisions are made, who makes them, who is aware of information, and how information is transferred; all of these factors play a major role in facilitating or hindering change. Finally, how individuals learn the institution's values and how they should behave, especially newcomers.

8.2 Definition of the Innovative University's Culture

Innovation is a combination of creativity, risk, and experimentation. Some universities have gained their stature when they managed to conduct successful experiments. But most analysts associated with higher education would probably say that the prevailing cultural norm is traditions, not innovation, whether the university is in Europe, Latin America, or the USA. However, the community needs good ideas to flourish. Universities need to be an incubator of such ideas. Firstly, let us get to know the innovation obstacles in universities, then have a look at the criteria that lead to innovation.

The obstacles hinder universities' creativity can be summed up as follows:

1. **Lack of motivation of change process and taking risks**
   If the university or institution reaps the benefits of their usual activities and affairs, there will not be a considerable motivation for change or trying and taking risks that change could pose. An author narrates an event he has experienced personally in one of big companies over 30 years ago, where the company was a public sector company (state company) and was importing a large-scale of die from a European country. At that time, the head of research and development sector found that the template price was very high, and he wanted the company engineers to gain the practical experience necessary to manufacture such templates, especially providing the necessary equipment and skilled technicians. So, he decided to dedicate two engineers to accomplish this task and gave them the authority of ordering raw materials necessary to carry out the initial experiments. Subsequently, the production of the template was accomplished in a six-month period after several failed design, manufacturing, and heat treatments experiments. When calculating the cost of all affairs relating to the first template (wages of engineers and technicians, manufacturing

processes, and the value of raw materials of all experiments), It was found to be less than the cost of the imported template, as well as how the two engineers had gained important experience and became experts in this matter. Unfortunately, three years later, another engineer headed the company's research and development sector. Manufacturing another part of template with volume did not exceed 1% of the previous one was discussed, but after controversy over, who would be responsible if the decision of manufacturing that part was taken and then it failed to perform its required function, and so forth, the decision was made to import and not to go through the manufacturing experience, despite there being a previous successful experience.

2. **Rules and regulations**

   In addition to the existence of rules and regulations, the institutions inform their personnel about inhibitions. This kind of instructions often discourage the creative aspect of personnel, as most institutions prefer that all individuals work according to the pre-established instructions and rules. As known, the more the bureaucratic rules and regulations there are, the less the possibility of the institution's members being innovative, hence the institution will never be innovative.

3. **Permanent system of monitoring individuals**

   Although evaluation is vital to any institution, a permanent monitoring system, especially for individuals, reduces creativity. Therefore, the evaluation system should enhance performance rather than monitor individuals for their irregularities or shortcomings, because when evaluation aims at improvement rather than punishment, creativity of individuals develops.

8.3 Conditions of Innovation Culture

To create and enhance an innovative environment, the university needs to develop a culture of innovation in individuals. Thus, seven conditions have been drafted to ensure that the university has a culture of innovation:

8.3.1 Motivation and Risk Culture

According to what is stated in reference [115], "Gerard Tellis" refers to "the deep cultural characteristics" existing in the stable organizations, which put obstacles in the way of change in fear of risks, such as the lack of reward for innovation and the severity of punishment for failure. As a result, innovators and potential innovators are afraid of failure. So, it is necessary to empower the individuals to utilize their skills optimally. Supervisors should recognize

individuals and their skills and create supporting environment thereof. Hence, the concept of supporting environment will transfer, in the institution, from one individual to another so that innovation culture prevails in all active actors of institution and so the concept of rewards related to experiment deepens.

8.3.2 Individuals Freedom of Controlling Means to Reach the Goal

Institution should encourage the individuals to control the means to reach the agreed goal, as the personnel in an innovative institution need some kind of independence. Thus, universities need to prepare the ground for enabling the pioneers and innovators do their jobs with complete ease and support. Support is not only financial rewards but also a culture considers risks positively.

8.3.3 Stability of Institution's Goals

If we would like to empower individuals to think creatively, the institution's goals should not be changed day after day, when institution's officials are substituted for example. As, for example, when the president of university says: Community participation is the most significant and, four years later, a new president says: Scientific research is the ultimate objective; a confusion would be created among personnel. Or when the president of university says: Teaching is the most significant, while the vice president says: Research is the most significant. This disrupts the individual culture, resulting in their failure to manage their work optimally. Therefore, when the institution is committed to an organized activity in a specific track, the system of success will be clear for all, and individuals will know what they must do to succeed. But in case if goals keep changing, individuals become less interested in sharing creativity or taking the risk of introducing creative ideas.

8.3.4 Prevalence of the Culture of Academic Freedom Among Individuals

The culture of institution should adopt the principle that the individual is of a close relevance and importance to what is happening in his institution in order to enable individuals invest in their environment, care about what is going on in their institution, and be interested in its excellence. Individuals need to be aware of where the institution is headed, all with a degree of independence.

This culture: Strategic orientation and personal independence give individuals a sense of responsibility and motivate them to do their best to achieve the due goals. This environment is utterly different from the environment of "production lines" where individuals work as if they are copies of each other.

8.3.5 Innovative Culture with Financial Resources and Time

Determining a task or a goal without providing the financial support means sending contradictory messages to individuals. While resources allocation represents a strong sign referring only to what is important. So, an institution having an "innovation fund" allows good ideas to move forward, because this fund represents a strong signal of the seriousness of institution. On the other hand, if the institution punishes individuals who are seeking to get external fund to carry out their innovative business, it is probably that innovation is considered as minor or unnecessary [115].

Time is another kind of significant resources. It can stimulate innovation or do the opposite. Researches show that when individuals work constantly under the stress of deadlines, they do not look for innovative solutions, but simply choose a way enabling them to meet the deadline. Then, if innovation is significant, the institution's way of thinking regarding how the individuals spend their time should be taken into consideration, as well as helping them in allocating their time between teaching, research, innovation, and creative activities.

8.3.6 The Institution Should Have the Culture of Collective or Joint Thinking

The dilemma of academic environments is that professors are often isolated or introverted, each of them prefer working on their own, while the "academic community" must be a complex group. So, if we want to create an innovation-oriented organization, we must pay close attention to "the culture of collective work or thinking," as the different experiences, several ways of thinking and varying ranges of age enrich the environment and make it innovative.

Collective vision comes through a variety of perspectives, which can produce an exciting thing. Diversity means the variation in viewing, classifying, understanding, and developing people toward improving the world. Therefore, the institution needs, on one hand, to pave the way for several perspectives and ideas, and, on the other hand, to be able to organize those visions into a coherent unit as players of a football team need to cooperate to achieve their goal. As far as regulation is concerned, if individuals do not gather

around a certain vision, commitment to innovation decreases. Also, if we do not respect what the individual offers to the team, the institution will end up with a culture of isolation and each individual will follow his own way.

8.3.7 The Moral Support for Institution Individuals

Gestural and financial resources are equally important. An innovative institution needs to create a culture of experiencing and risk taking, i.e., the leadership of an institution has to provide moral support to creative people. To create a culture for innovation in a university, individuals in higher education organizations shall be trained in the art of critical thinking. In addition, it is necessary to organize group discussions to present the findings of individuals and teams, as such discussions enrich the work, create ideas, uplifting creativity, and establishing the culture of innovation. But if these groups are turned into a way to criticize others and devalue what they have done, it would result in narrowing down creativity and indicate individuals' flawed understanding of the importance of debate with scientific methodology.

8.4 Factors That Make the University Innovative

The scientific research and universities' innovations have become of universities evaluation criteria on a global scale. The true indicator of the progress of countries and the advancement of their communities is excellence in terms of innovation. Accordingly, scientific research is no longer a luxury for educational institutions in order to reach creativity and excellence to achieve sustainable development, it rather became a necessity for its ability to solve many economic, health, educational, and social problems, according to proper scientific bases. In light of that, it was only natural for universities to pay attention and direct their activities to train their cadre of researchers and students to master the scientific research methods that lead to creativity and subsequently innovate new products or develop existing products resulting in the advancement and prosperity of community.

To be innovative, the top management of university should create and promote number of factors and conditions that help in developing the innovation culture among its individuals, leading to be one of the worldwide innovative universities. These conditions and factor can be summed up as follows:

8.4.1 Spreading Innovation Culture

Universities need to incubate good and new ideas that create innovation, helping in the prosperity of community. If innovation is a combination of creativity, risk, and experimentation, then some universities have gained their stature and even pioneering when they managed to conduct successful experiments that offered distinctive products. To be innovative, the culture of creativity should prevail in the university. So, it is necessary for the university to spread innovation culture among its personnel and work on encouraging creativity and creating innovative environment.

8.4.2 Collaboration with Industry

The importance of universities' role in economic and social development is obvious in that it is not limited to educational process, but it extends to the scientific research that results in innovations and inventions. These innovations and inventions are perceived as an outcome of knowledge transferring and marketing among the different industries and technologies. This results in a collaboration between the university and the various industrial sectors, and universities effectively contribute to the economic development. In addition, community gains multiple benefits through this close collaboration between the universities and industry. The importance of collaboration between universities and business organizations and establishing an active partnership between them is emphasized through the universities' contribution to developing the performance of industrial institutions and business organizations on one hand; on the other hand, promoting the universities' competitive ability as well as achieving the required quality in their programs and outputs. The need to promote links between universities and private sector has notably increased because of universities' need to develop their financial resources of the private sector, as they represent an alternate source of financing the institutions of higher education. Also, due to institutions' need to the science, innovations, and advanced services of universities. The world's universities and companies are making great efforts to develop their economic situation and increase their competitive abilities and interest in keeping up with the technological development and scientific innovations.

Here are some examples of innovative universities and their collaboration with their countries' industry:

1. In the fiscal year 2016, the expenditure on researches in MIT amounted to more than $728 million; 19% of which came as support by industry, and there are about 700 companies that works with teaching staff and students.

2. In 2016, the University of KU Leuven, Belgium, announced a new partnership with Ford Motor company to examine the durability of 3D Printed cars' parts.
3. In the fiscal year 2014/2015, the budget of supported researches in Harvard University reached $800 million; $50 million of which were collected from the companies' support.
4. The expenditure on researches in University of Texas exceeds $2.7 billion; $641 million of which are collected from the private sector resources.
5. University of Waterloo, Canada, has been working alongside industry since its founding with industry and allowing people to own their intellectual property and benefit from the income of marketing, as the university has gathered millions of dollars for researches from donor agencies and industry. Researches were funded with $183 million by institutions, industry, and non-profit organizations in (2015/2016).

8.4.3 Contributing to Creating Start-Ups and Incubators

The results of researches come first in the innovation process, but research alone neither drives the economic growth nor achieves prosperity. Systems should be put in place to drive and support innovation so that it leads to licensing new products and/or emergence of new subsidiaries of the university. This must include teams of experts to encourage business activities and consultations and increase interaction with the industry.

Also, world's prestigious universities need systems and devoted people to interact with the business outside their walls to provide advice and expertise and support business partnerships. These are specialized fields, as individuals or work teams specialized in this field majorly benefit from external partnerships, as they are the nexus of interaction with the business world and companies. Such teams can become extraordinarily important in improving the university's characteristics and reputation in the business world on globally. Additionally, we note that specialized teams in this field are very useful in guiding the academic medium in the labyrinths of this unfamiliar world, significantly helping to manage risks and preserve the image of the university.

Start-ups must be helped to get funded from selling shares. This purpose is served by inviting large companies to establish partnership to support start-ups by providing incubation, joint funding, and guidance, as well as the possibility of giving them priority to purchase their products. Such program increases the entrepreneur's chances of surviving in the market.

Since research and development investments yield revenues on the medium or long run rather than short run, providing financial support may be a concern for start-ups. Lately, this concern has resulted in designing programs that support or create venture capital funds. Venture capital (or risk capital) is a type of funding entrepreneurship projects in its early stages, in which it promises high potential of success and growth. However, investing in such projects is of high risk. Venture capitalists gain revenues from their shares of the project they invest in, which usually has new technology or action plan in case of high-tech companies such as: Biotechnology, information technology, software, artificial intelligence, electrical engineering, information systems, photonics, nanotechnology, nuclear physics, robots, wired and wireless communications, and automobiles. Venture capital is fundamentally different than credit or loan. In case of loans, the creditor may claim their money regardless of the company's condition or financial stability; whereas, the venture capitalist invests in a company and gets a share of it, which means that his revenues depend entirely on the project's growth and its ability of making profits. The institutions of venture capital are considered a means of financial, technical, and managerial support for start-ups of high risk and high growth and profit potentials. Examples of universities classified among aforementioned pioneering, innovative universities show how much interest these universities have for contributing to the establishment of many start-ups by their teaching staff or alumni. Here are some examples:

1. At University of Pennsylvania in 2016, 22 start-ups were established by business incubators and innovation center of the university; yielding revenue of over $50 million.
2. In 2016, 21 new start-ups were launched at University of Washington, bringing the total number of start-ups within 10 years to 126.
3. As for University of California, until February 2018, 1029 start-ups were established.
4. The Swiss Federal Institute of Technology in Lausanne has an innovation park which hosts laboratories, offices, and researchers from more than 150 start-ups and operating companies. In 2017 only, the school initiated 15 start-ups, and 115 companies in the Innovation Park. The start-ups have raised their budget to 112 million Swiss francs.
5. The University of Waterloo has the biggest incubator that provides over 1200 jobs and $425 million by velocity companies.

8.4.4 Supporting and Motivating Personnel

Support and motivation are the main core of any successful management and they are the driving force of all institutions to move toward achieving targets. Motivation is defined as a set of factors that influence the will and motives of individuals in universities and companies to achieve these goals. There is no doubt that the provision of material and moral support to people is necessary to stimulate their creative thinking, especially the creative ones among them. Some companies such as (Walt Disney), the global company, embody the principle that (great ideas can be inspired by anyone). This is why the company offers money for any idea that can be turned into a product to encourage innovation and increase its productivity. Creating proper environment for the employee, motivating any outstanding work he does, and acknowledging his achievements as being part of the institution drives the employee's motivation over time to come from within, which, in turn, increases his trust and loyalty to his belonging to the institution, as well as developing creative thinking.

8.4.5 Establishing a Firm for Marketing the Universities' Products

In the University of Waterloo, Canada, there is a business management office that markets the university's products. In 2018, it was marketing for 275 patents.

8.4.6 Raising Funds for Applied Researches

Endeavoring to obtain funding resources for outstanding researches is an excellent way to promote scientific research and support research programs. It also encourages the necessity of keeping relevant to methodological progress that contributes to upgrading the quality of research activity. Reviewing examples of universities classified among pioneering innovative universities mentioned in this book shows how much interest these universities have in collaborating with the private or state business sectors to bring huge funds used for conducting researches; here are some examples:

1. Texas's universities spend over $2.7 billion; $641 million of which are obtained from private sector sources.
2. In the fiscal year 2016/2017, Stanford University announced a research budget of $1.6 billion, as it has over 6000 projects supported from other entities.
3. As for the fiscal year 2014/2015, the budget of supported researches in Harvard University was nearly $800 million; $50 million of which were collected from the companies' support.
4. In July 2015, Defense Advanced Research Projects Agency (DARPA) provided $21.6 million for the scientists of University of California, Berkeley to develop a

"brain modem" that could directly stimulate thousands of neurons using expected light.

5. Northwestern University received $649.7 million for funding researches within the 2015/2016 academic year.
6. The revenues of Pohang University of Science & Technology, POSTECH, from scholarships and research contracts exceeded $149 million in fiscal year 2015.
7. In 2016, the revenues of California Institute of Technology reached $2.6 billion, 46.5% of this budget is scholarships and research contracts.

8.4.7 Reducing the Student–Faculty Members Ratio

Below are some numbers relating to universities ranked among the world's pioneering innovative universities mentioned in this book and the student–faculty ratio. This ratio indicates the quality of interaction among them as the number of students per faculty member becomes fewer; the results of the quality of the interaction is also reflected on the quality of learning and scientific research:

1. Number of students in Duke University is 15,192, and number of academic teaching staff members is 5022; which means that the student–faculty ratio is 3:1
2. In California Institute of Technology, the student–faculty ratio is 3:1
3. The students' number in Northwestern University is 21,823, while the academic staff members are 3313; which means that the student–faculty ratio is less than 7:1.

8.4.8 Research Budgets

Here are examples of universities ranked among the pioneering innovative universities mentioned in this book. They show how great research budgets are, which helped a lot, along with other factors, to upgrade the quality of research and achieve creativity and innovation:

1. In the fiscal year 2017, the budget of scientific research in Pennsylvania State University amounted to $928 million, while in 2016, it exceeded a billion dollars for granting the sponsored projects.
2. In the fiscal year 2015, the total research expenditure of KU Leuven—Belgium exceeded €454 million.
3. In the fiscal year 2016, the research budget in University of Washington amounted to $995 million, equivalent to 72% of the university's annual budget.

4. University of Michigan spends annually over $1.3 billion on researches.
5. In the fiscal year 2016/2017, Stanford University announced a research budget of $1.6 billion.
6. Texas universities spend over $2.7 billion on researches.
7. In the fiscal year 2016, the total research expenditure of Vanderbilt University—USA was approximately $235 million on the scientific research, and it supported projects and researches with a total of $214 million.
8. In 2015, University of California received over $4.97 billion to fund its researches.
9. Northwestern University received $649.7 million for funding researches during the 2015/2016 academic year.
10. The budget of Pohang University of Science & Technology, POSTECH for 2016 reached $304 million, $141 million of which is were allocated to research, which accounted for 46% of the overall budget.
11. From 2013 to 2017, the annual grants of research in Imperial College ranged between £330 and 436 million annually. In addition to the university's expenditures for research and equipment out of its own budget, which almost amounted to £1 billion in 2017.
12. In 2016, the revenues of California Institute of Technology reached $2.6 billion, 46.5% of this budget is scholarships and research contracts.

8.4.9 Other Factors

In addition to the previous eight main points, there are a number of factors that drive the university to be innovative. They have been deduced from the research supported by King Abdul Aziz University [116], summarized as follows:

1. Encouraging the research endeavor using cross-disciplinary method; which means forming research teams of different disciplines, as it is more creative than the other methods.
2. Forming a professional team to provide support for the identification, protection, and utilization of intellectual property by providing investors capable of supporting it financially, and to turn ideas into products of commercial value.
3. Linking the course of scientific researches to community's problems and needs and the possibility of identifying the projects of research group in consultation with the beneficiaries.
4. Linking the university education and the postgraduate researches to the national companies' requirements to enable the university to become magnet for the best talents.

5. Developing rules and regulations to support and consolidate the cross-disciplinary research, allowing the creation of research teams from several fields for addressing complex global and community problems.

8.5 The Kingdom Vision 2030 and Innovation

8.5.1 Introduction

Allah has granted the Kingdom of Saudi Arabia many geographic, civilizational, social, demographic, and economic characteristics that enable it to occupy an esteemed position among leading countries. The vision of any country's future starts from its strengths, and this is what has been considered when shaping a vision for the Kingdom since 1452 H (2030). The Kingdom position in the Islamic world will enable it deeply and supportively to play a leading role in the Arab and Islamic nations, and its investment power will be the key and motive to diversify the economy and achieve sustainability. Also, the Kingdom strategic location will enable it be a hub for the three continents.

The Kingdom Vision 2030 is based on three pillars shown as follows [117]:

1. Vibrant Society,
2. Thriving Economy,
3. Ambitious Nation.

These pillars are integrated and consistent with each other in order to achieve the desired goals and maximize the benefit from this vision. The first pillar (vibrant society) makes it clear that the community lies in the heart of the Kingdom's vision and is considered the basis for achieving this vision and establishing a strong foundation for the Kingdom economic prosperity. This pillar comes from the belief in the importance of building a vibrant society where its members live according to the Islamic principles and the approach of moderation, while being proud of their national identity and their ancient cultural heritage. The vision also aims to achieve a positive and attractive environment, where constituents of life quality for citizens and residents are provided. Moreover, the vision aspires for a coherent family structure and empowered social and healthcare systems.

The vision's second pillar (thriving economy) focuses on providing opportunities for everyone by building a proper education system aligned with labor market needs and developing opportunities for entrepreneurs, small-sized enterprises as well as large companies. The vision also focuses on developing investment tools to unleash the potential of promising economic sectors, diversify the economy and create new job opportunities for citizens. In addition, the vision aims to concentrating the efforts on privatizing the government services and improving the business environment. This will contribute to attract the finest international talents and the best qualitative investments and make use of the Kingdom's unique strategic location.

The vision's third pillar (ambitious nation) focuses on defining the role of an effective government that fosters efficiency, transparency, and accountability. This likely to encourage the culture of performance in order to enable the Kingdom's human resources; create the necessary environment for citizens, business sector, and non-profit sector that have them bear responsibilities; and take the initiative to face challenges and seize opportunities. In this concern, the vision states that "to be among the best 20–30 educational systems in coming years worldwide."

8.5.2 Vision 2030, Universities and the Kingdom's Rank on Global Competitiveness Classification Index

The Global Competitiveness Report (GCR) is an annual report that is issued by the World Economic Forum since 2004. It classifies 140 countries according to the global competitiveness standard, which is the internationally recognized ranking. In October 2018, the last report was issued where the countries' ability to provide their citizens with economic welfare and prosperity were assessed. This is in turn depends on the country's ability to benefit from its available sources. Therefore, the global competitiveness scale evaluates the economic ability of every country to achieve sustainable growth on the medium and long term. This scale depends on 12 pivots or main pillars, each of which consists of a number of indicators, with a total 98. The report focuses on the changing nature of economic competitive ability in an increasingly changing world by new digital technologies, which introduce a new group of challenges for governments and companies.

The interest of countries and national institutions in competitiveness has increased for several reasons on top of which is to join the local and global market competition. This reflects the country's ability to achieve high growth rate and improve living standards for its citizens, as well as its appropriateness for foreign investments. The countries ranking, on competitiveness scale, can be significantly improved through higher education, continuous training, developing staff abilities and skills, in addition to making use of technology that helps to improve products quality.

In 2018, the Kingdom was ranked the 39th worldwide. Although all pillars and indicators counted for global competitiveness index are considerably based on human resources such as institutions, health care, economy, technological readiness, etc., universities play a key and direct role in the progress of any country when it is related to skills and innovation in particular. Among its interests, Vision 2030 focuses on improving the Kingdom's rank in competitiveness index. An article about the Kingdom's objectives by 2030 states: "Reaching one of first ten ranks in global competitiveness index," along with the focus of Vision 2030 on pillars of innovation and skills. A number of goals show the interest in innovation and education (the main responsible factors for skills). The main objectives of the Kingdom's Vision 2030 includes, for example: "Increasing the employment rates," from which the following sub objectives are derived:

1. Developing human capital according to job market needs, from which a number of detailed objectives are derived, such as:
2. Improving the basic education outputs,
3. Improving the ranking of educational institutions (such as universities),
4. Providing qualitative knowledge for distinctive individuals in fields of priority,
5. Expanding the vocational training to meet the job market needs.

Enabling the creation of job opportunities through small and medium-sized enterprises, as well as micro-projects. A number of detailed objectives are derived from this objective, such as:

1. Enabling and supporting the culture of innovation and entrepreneurship,
2. Increasing the contribution of small and medium-sized enterprises to the economy,
3. Increasing the contribution of productive families to the economy,
4. Attracting global talent that best suit the economy.

8.5.3   Extracts from the Kingdom's Vision 2030 Showing Its Interest in Innovation and Skills

Privatizing Government Services
We believe in the role of the private sector, and therefore, we will open up new investment opportunities in order to encourage innovation and competition, and we will remove all obstacles preventing the private sector from playing a larger role in development. We will continue to develop and activate the legislative system of markets and business. This facilitates greater opportunities for investors and enable the private sector to provide some services in the sectors of health care, education, and so forth. We will seek to shift the government's role from «a service provider» to «an organizer and monitor of sectors». We will also prepare the necessary capabilities for monitoring the level of services in the concerned authorities. As the private sector currently contributes with less than 40% of GDP, we will work on increasing this contribution by encouraging local and foreign investment in sectors of health care, municipal services, housing, finance, energy, and so forth. All of this will be subject to a flexible management and effective control.

Among Our Commitments: Bigger Role for Small and Medium-Sized Enterprises
Small and medium-sized enterprises (SMEs) are among the most important drivers of economic growth, as they work on creating job opportunities, supporting innovation, and boosting exporting. SMEs contribute with low percentage of GDP compared to advanced economies. We will seek to create suitable job opportunities for citizens all over the Kingdom by supporting entrepreneurship, privatization, and investment programs in new industries. In this respect, we established the general small and medium enterprises (SME) authority. We will continue to encourage young entrepreneurs to achieve success through establishing better regulations and by-laws, facilitating access to funding, increasing international partnerships, and allocating a greater share of governmental procurements and bids. We will support productive families that modern means of communications allowed wide marketing opportunities for them. This is through facilitating opportunities for funding micro-projects and motivating the non-profit sector to build the capabilities of those families and fund their initiatives.

SMEs contribute with only 20% of GDP, compared to the percentage achieved in advanced economies that can reach up to 70%. Despite the efforts exerted to improve the level of business environment, SMEs in the Kingdom still endure slow and complex legal and administrative procedures. They also struggle to attract talents and get access to funding, as the percentage of SME funding is no more than 5% of the overall funding—a far lower percentage than the global average. We will strive to facilitate access to funding and to encourage our financial institutions to allocate up to 20% of overall funding to SMEs by 1452 AH (2030 AD).

The recently established SME authority will seek to review regulations and by-laws thoroughly, remove obstacles, facilitate access to funding, and enable the youth and innovators to market their ideas and products. At the same time, we will seek to establish more new business incubators, specialized training institutions, and venture capital funds. These will aid entrepreneurs in developing their skills and innovative skills. We will also help national SMEs in exporting and marketing their products and services, by supporting e-commerce and collaborating with international stakeholders.

We Learn to Earn

We will continue investing in education and training, providing our youth with the necessary knowledge and skills for their future jobs. Our objective is providing every Saudi child, wherever he/she lives, with higher quality education opportunities based on various options. We will focus particularly on developing early childhood education stages, preparing our teachers and educational leaders, and refining our educational curriculum.

We will also increase our efforts to ensure that the outcomes of our education system are in line with job market needs. As we have launched the National Labor Gateway (Taqat), and we will establish sector-specific vocational councils that will precisely determine the skills and knowledge required for each developmental sector. We will also expand vocational training in order to push forward economic development. Our scholarship opportunities will be directed to prestigious international universities and be awarded in the fields that serve our national priorities. We will also focus on innovation in advanced technologies and entrepreneurship.

"We Will Focus on Innovation in Advanced Technologies and Entrepreneurship"

An important quote on supporting innovation in advanced technologies and entrepreneurship, as focusing on this field is what these developed countries seek to.

8.6 King Abdulaziz University and the Efforts to Acquire Patents

Innovation is the distinguished characteristic of the speed of university research development, and the following transferring the components of technology to community. There is no doubt that registered patents indicate the high position of the university and researchers. Therefore, the university spares no effort to help researchers acquire patents for their innovations and support them in all what it takes to achieve this noble aim. In the last three years (2017–2019), globally registered patents of King Abdulaziz University was successively achieved and the fruits of labor were as follows:

1. Fida Needle: A medical instrument carves body cartilages without any need for surgical intervention. It was launched in global markets in its first version under the name of "Fida Needle" at the beginning of 2017 in Global Health Exhibition in Dubai, after the journey of this invention was sponsored by King Abdulaziz

University since its launch. This invention was implemented by a team from the university consisted of the inventor Meshal Hisham Al-Harasani, Dr. Abdul Karim Reda Fida, and Dr. Faisal Hazem Zaqzouk, after the team got a globally registered patent issued by King Abdulaziz City for Science and Technology. In order to turn the idea into a physical reality, It took 6 years of work, research, studies, and participation in several conferences. The Saudi French invention also went beyond the European technologies in the same field. This is after using it in the Germen capital Berlin at Helios hospitals for the first surgery of carving children ear cartilages and modify its hillock with the minimum surgical intervention and cost. This invention is considered one of King Abdulaziz University accomplishments and its contributions in achieving the Kingdom Vision 2030 in the fields of medicine, innovation, and knowledge economy. This invention has the logos of the Kingdom's Vision 2030 and King Abdulaziz University on all global products, as it is shown in Fig. 8.1.

2. A team of experts in the anatomy department in the faculty of medicine consisting of Eng. Yahia Badr, Dr. Abdel Mounem Hayani, and Dr. Mohammad Ba-Dawoud registered a patent for a new and innovative global level method for maintaining autopsies and biopsies used in teaching students the human body formation by utilizing a method using "Shellac"— rather than the harmful "Formalin".

3. Dr. Ramzi Reda Obaid, member of teaching staff of the Electrical Engineering and Computer Engineering Department at the Faculty of Engineering, registered two patents in the USA on mechanical errors in induction electrical motors. The researcher managed to develop and design two systems through a scientific research for Eaton Corporation in the States of Wisconsin and Ohio in the USA.

4. Prof. Yahia Abu Bakr Al-Hamid, Prof. Abdul Rahim bin Ahmed Al Zahrani, and Dr. Mohammad bin Abdul Rahim Daous, members of teaching staff of the Chemical Engineering and Materials Department at the Faculty of Engineering, got a patent in the field of deoxygenation operations with Luria-alumina catalyst. They got it from European Patent Office in Munich, Germany. This invention is concerned with using this dehydrogenation to turn Ethyl benzene into Alastrin as a highly important aromatic hydrocarbonic substance, widely used as a raw material for producing synthetic rubber.

5. Dr. Abdulrahman Hamed Abdullah Al-Massoud, a member of teaching staff of the Electrical Engineering and Computer Engineering Department at the Faculty of Engineering, came up with two inventions in the

Fig. 8.1 Cover of "Fida Needle" invention

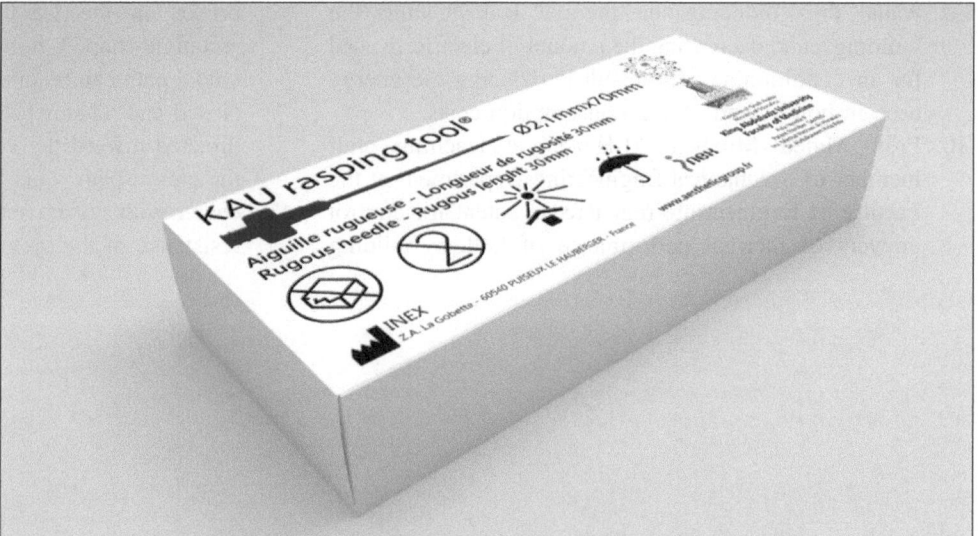

field of electrical group. He registered them in King Abdulaziz City for Science and Technology. The first invention is entitled "Neutral Point Technology for Generator Protection" and the second is entitled "Zinc and Copper Circuit Breaker."

6. A research team consisted of Prof. Fahad Al-Marzouqi and Prof. Adnan Zahed (Chemical Engineering and Materials Engineering Department at the Faculty of Engineering), Prof. Antonio Bezzi and Dr. Sayed Abdullah (Physics Department at the Faculty of Science), registered a patent in the USA for producing a new material that can be used for manufacturing car brakes from bio-colloids. The new material is a mixture of ground agricultural wastes, alcohol, natural materials, and some oxidants, all molded in the desired shapes.

7. Prof. Majed Muallah Al-Hazmi, a teaching staff member of the Mechanical Engineering Department at the Faculty of Engineering, registered a patent for flash evaporation station equipped with feeding water cooler, in the USA. The system consists of a regular flash evaporation station equipped with a cooling unit, where the water passes firstly through the cooling unit, which reduces water temperature to the last stage of evaporation stages. This leads to an increase in the evaporative range—thus increasing pure water rates. Both flash evaporation station and the cooling unit are already known regular systems. However, the point is using them in an integral way to increase the evaporation range in addition to making pure water production rates unaffected by climatic changes.

8. Prof. Majed Muallah Al-Hazmi, a teaching member staff of Mechanical Engineering Department at the Faculty of Engineering, registered a patent for a desalination plant that operates by hydrogen fuel, in the USA. MSF or MED desalination plant uses hydrogen fuel to operate the top saltwater heater in the factory. Combusting hydrogen fuel with oxygen leads to generate carbon-free hot gases. This combustion of hydrogen fuel and oxygen occurs under liquid water spray within a directly connected room. The vapor resulted is used to heat animal feed to reach the top saltwater heater and increase its temperature up to the level needed to start evaporation. Hydrogen is produced from seawater with solar energy. In order to operate the desalination plant continuously, a suitable hydrogen storage system can be added to the hydrogen production unit. It is suggested to use the hydrogen, produced from seawater by solar energy, to generate the vapor necessary to run the thermal desalination process.

9. Prof. Majed Muallah Al-Hazmi, a teaching staff member of Mechanical Engineering Department at the Faculty of Engineering, registered a patent for a thermal inhibitor of the hollow building bricks, in the USA. The invention is a chip made of materials of relatively low thermal conductivity, folded in an inverted G-letter shape, with straight and sharp edges. After being folded, the chip is inserted into the hollows of the hollow bricks. The chip is folded with a larger height than the mid-height of the hollows in hollow bricks, in order to form a narrow path that restricts the movement of natural thermal convection currents caused by the temperature difference between the sides of the hollow building bricks. Reducing the thermal convection in the hollows of bricks increases the heat resistance of the

walls, thus reducing the thermal leakage into the buildings, and decreasing the amount of electricity used for air conditioning. The result is reducing the energy used in houses and saving consumption costs.

10. Prof. Majed Muallah Al-Hazmi, a teaching staff member of Mechanical Engineering Department at the Faculty of Engineering, registered a patent for control drawers of thermal performance of hollow building bricks, in the USA. A building brick contains a rectangle-shaped hollow where a drawer with helicoidal path can be inserted. The helicoidal path forms a closed-end channel that can receive hot air that is directed toward the central area of the internal end of the closed path. This leads the air to accumulate in the center with no movement, thus increasing the thermal resistance of the system.

At the end of this book, we conclude that innovative universities should work in different ways to reach the classification of innovative universities. To become among the innovative universities, our university (KAU) depends on some factors that can be summarized as follows:

1. Spreading the culture of innovation among the academic staff and university students especially the postgraduates,
2. Collaborating with the sector of industry, particularly the industry near the surrounding area of the university, as well as national industries in general,
3. Contributing to building start-ups and incubators, because they are more likely to achieve innovations,
4. Supporting and motivating staff, whether academic or non-academic, as well as senior and postgraduate students,
5. Opening a university office with marketing experts whose main task is to market the products of the university,
6. Getting funds for applied research, from different internal and external authorities,
7. Reducing the gap between numbers of students and teaching staff, in order to focus more on productivity,
8. Increasing and diversifying the sources of research budget in the university,
9. Encouraging interdisciplinary research by creating research teams that involve different disciplines, a more creative way than specialized teams,
10. Supporting cross-disciplinary research in several areas to address complex local and international problems,
11. Defining, protecting, and benefiting from intellectual property by communicating with investors who are able to move from ideas into products of commercial value,
12. Taking the course of scientific research to address the problems and needs of the community and beneficiaries,

13. Correlating university education and research to the requirements of local companies so that the university could attract the best talents,
14. Supporting student excellence, as leading universities in the late twentieth century realized that the absence of academic differences makes students lose ambition. Therefore, universities have started to look for new ways to distinguish hardworking and excellent students. For example, Harvard University adopted a major competitive sports championship throughout the semester. Students became enthusiastic and sought to achieve excellence in sports. Thus, sports have become a source of motivation that highlights competition in classrooms, where students are getting ready for practical life. The reinforcement system that encourages incentives and supports for academic excellence has continued to achieving satisfactory results in general [20].
15. Selecting teaching staff and employees based on competency criteria, for the appointment of outstanding teaching staff, is the most important decision taken in the university. It is also necessary to enhance the diversity of teaching staff and students.
16. Creating educational programs and research centers that promote creativity and innovation, while evaluating capabilities and selecting appropriate options,
17. Competitive facts indicate the need of most traditional universities to restructure and identify themselves. A university should start this by assessing its most valuable assets including: teaching staff and infrastructure. Taking these assets in consideration, the following question arises: How good is the university's position? Not only compared to others, but also in terms of meeting the needs of students and other parties, the university supports discovery, creativity, and innovation.
18. Adopting the concept of "innovation is a strategy for success," i.e., the university policy is to work, progress,

© The Author(s) 2021
The Leading Worldâs Most Innovative Universities,
https://doi.org/10.1007/978-3-030-59694-1_9

and succeed, not to take precedence. This is because precedence is a result of progress and success.

19. Offering university free courses to outstanding school students in order to make them aware of the creativity achieved around the world, and what benefits individuals and communities can get. Here, instructors of such courses should believe in this importance, maintain interesting presentation techniques, and discuss topics of interest with students in such age.

20. Assigning instructors who show the information that encourages creativity and use interesting knowledge transfer techniques with pre-university students, thorough paying visits to schools to deliver lectures on the subject,

21. Opening the door for supporting any research conducted by schools, by allowing school students to access the university equipment for their research. This will not only lead to the success of scientific research and encouragement of creative researchers, but also to

establish a personal connection between the university and the student who will always remember the place that supported him/her.

22. Admitting students who ranked first in the department to postgraduate studies once graduated, without having to go through the routine process of application that may take more than a year, and make students lose enthusiasm, or probably apply for other jobs,

23. Encouraging the outstanding university students through well-planned propaganda and a committee that supports the participation of students in international competitions,

24. Including creative postgraduate students into the teaching staff once they receive their doctorate degrees. Research papers of such students are often not only for scientific promotion, but also for creativity that solves community problems or helps industrial and economic development.

References

1. Badran, I. (2004), The Role of Intellectuals in Establishing the Knowledge Society, 1st Arabic Conference on Human Development, Manama, Bahrain, 8–9 December 2004. (In Arabic).
2. Sternberg, R. J. and Lubart, T. I. (1999). The concept of creativity: Prospects and paradigms. In R. J. Sternberg (Ed.), Handbook of creativity (pp. 315–). Cambridge: Cambridge University Press.
3. Craft, A. (2005). Creativity in schools : tensions and dilemmas. London: Routledge.
4. Abdulwahed, M. N. and Diab, A., Fundamental Introduction to the Knowledge Society, 9th Conference of the Arab Higher Education and Scientific Research Ministers, Damascus, Syria, 15–18 December 2003. (In Arabic).
5. West, M. A. and Richards, T. (1999). Innovation. In M. A. Runco and S.R. Pritzker (Eds.), Encyclopedia of creativity (pp. 4556). San Diego, Calif.; London: Academic Press.
6. Taylor, C. W. (1988). Various Approaches to and Definitions of Creativity. In R. Sternberg (Ed.), The Nature of Creativity: Contemporary Psychological Perspectives (pp. 99- 121). New York: Cambridge University Press.
7. Sternberg, R. J. and Lubart, T. I. (1999). The concept of creativity: Prospects and paradigms. In R. J. Sternberg (Ed.), Handbook of creativity (pp. 315). Cambridge: Cambridge University Press.
8. Cropley, A. and Cropley, D. (2007). Using Assessment to Foster Creativity. In A.-G. Tan (Ed.), Creativity. A Handbook for teachers (pp. 209–230). Singapore: World scientific.
9. Zubair, M. and Shaoki, J., Investment in the Intellectual Capital as a Gateway to achieve Competitive Excellence, 5th Conference on Intellectual Capital in the Arab Business Organizations in the Shade of Modern Economies, Hasiba Bin Bou Ali University, Algeria, 13–14 December 2011. (In Arabic).
10. Klenow, Peter and Andrés Rodríguez-Clare.(1997), "Has the Neoclassical Revolution Gone Too Far?" NBER Macroeconomics Annual 1997, Brookings Institution.
11. Bahauddin, H. K. Education and the Future, Dar Al-Maaref, Cairo, 1997. (In Arabic).
12. Al-Ziadat, M. A., New Trends in Knowledge Management, Dar Safa'a for Publishing and Distribution, Amman, Jordan, 2008. (In Arabic).
13. Al-Shar'ee, B. G., The Role of Universities in Knowledge Creation: the Present Status and the Future, 6th Conference on Management, Salala, Oman, 10–14 December 2005. (In Arabic).
14. Al-Khedairy, M. A., Creation of Competitiveness Excellence: a Procedure for Achieving Advancement through Sustainable Development, Nile Group Publishing, Cairo, Egypt, 2001. (In Arabic).
15. https://www.mep.gov.sa/ar/vision-2030.
16. https://en.wikipedia.org/wiki/List_of_colleges_and_universities_in_Massachusetts.
17. https://www.shanghairanking.com/.
18. https://www.topuniversities.com/.
19. https://www.timeshighereducation.com/world-university-rankings/.
20. Christensen, C.M and H.J. Eyring, The Innovative University: Changing the DNA of Higher Education from the Inside Out. Jossey-Bass, 2011. ISBN 978-1-118-06348-4.
21. Harding, A. et al, (Editors) (2007), Bright Satanic Mills: Universities, Regional Development and the Knowledge Economy: Surrey (UK), Ashgate Publishing.
22. Felali, E. Y., the Role of Technical Assembly in the Transfer to Knowledge Economy, Book 27, Strategic Studies Center, King Abdulaziz University, Jeddah, Saudi Arabia, 2010. (In Arabic).
23. https://ar.wikipedia.org/wiki.
24. Tayeb, O., A. Zahed, and J. Ritzen (eds), *Becoming a World-Class University: The case of King Abdulaziz University*. Cham, Springer, 2016. DOI https://doi.org/10.1007/978-3-319-26380-9, Print ISBN 978-3-319-26379-3, Online ISBN 978-3-319-26380-9.
25. https://www.reuters.com/innovative-universities-2018/methodology.
26. https://www.reuters.com/article/us-amers-reuters-ranking-innovative-univ/reuters-top-100-the-worlds-most-innovative-universities-2018-idUSKCN1ML0AZ.
27. https://www.prnewswire.com/news-releases/clarivate-analytics-data-powers-the-annual-reuters-ranking-of-the-worlds-most-innovative-universities-300729189.html.
28. https://www.reuters.com/article/us-emea-reuters-ranking-innovative-unive/europes-most-innovative-universities-2018-idUSKBN1HW0B4.
29. https://www.reuters.com/article/us-asiapac-reuters-ranking-innovative-un/asia-pacifics-most-innovative-universities-2017-idUSKBN18Y24R.
30. www.stanford.edu.
31. https://facts.stanford.edu/pdf/StanfordFacts_2017.pdf.
32. https://en.wikipedia.org/wiki/Stanford_University.
33. https://innovation-ranking.clarivate.com/innovation-ranking/landing.
34. web.mit.edu.
35. https://ocw.mit.edu/index.htm.
36. https://en.wikipedia.org/wiki/MIT_OpenCourseWare.
37. https://en.wikipedia.org/wiki/Massachusetts_Institute_of_Technology.
38. https://mitadmissions.org/apply/process/stats.
39. https://web.mit.edu/facts/faculty.html.

The Leading Worldâ s Most Innovative Universities,
https://doi.org/10.1007/978-3-030-59694-1_1

40. www.harvard.edu.
41. https://online-learning.harvard.edu/.
42. https://college.harvard.edu/admissions/admissions-statistics.
43. https://www.harvard.edu/media-relations/media-resources/quick-facts.
44. https://en.wikipedia.org/wiki/Harvard_University.
45. www.upenn.edu.
46. https://www.upenn.edu/about/facts.
47. https://en.wikipedia.org/wiki/University_of_Pennsylvania.
48. www.washington.edu.
49. https://en.wikipedia.org/wiki/University_of_Washington.
50. www.utsystem.edu.
51. https://en.wikipedia.org/wiki/University_of_Texas_System.
52. www.kuleuven.be.
53. https://en.wikipedia.org/wiki/KU_Leuven.
54. www.imperial.ac.uk.
55. https://www.imperial.ac.uk/about/introducing-imperial/facts-and-figures.
56. https://www.imperial.ac.uk/media/imperial-college/administration-and-support-services/finance/public/Annual-Report-and-Accounts-2017.pdf.
57. https://en.wikipedia.org/wiki/Imperial_College_London.
58. https://www.unc.edu/about/by-the-number/.
59. https://en.wikipedia.org/wiki/University_of_North_Carolina_at_Chapel_Hill.
60. www.vanderbilt.edu.
61. https://en.wikipedia.org/wiki/Vanderbilt_University A.
62. www.kaist.edu.
63. https://en.wikipedia.org/wiki/KAIST.
64. https:// www.epfl.ch.
65. https://information.epfl.ch/facts.
66. https://en.wikipedia.org/wiki/%C3%89cole_Polytechnique_F%C3%A9d%C3%A9rale_de_Lausanne.
67. www.postech.ac.kr.
68. https://www.postech.ac.kr/eng/about-postech/introduction-to-postech/status/.
69. https://www.universityofcalifornia.edu/uc-system.
70. https://www.universityofcalifornia.edu/press-room/fact-sheets.
71. https://www.universityofcalifornia.edu/sites/default/files/uc-at-a-glance-feb-2018-final.pdf.
72. https://www.usc.edu.
73. https://about.usc.edu/facts.
74. https://en.wikipedia.org/wiki/University_of_Southern_California.
75. https://: www.cornell.edu.
76. https://www.cornell.edu/about/facts.cfm.
77. https://en.wikipedia.org/wiki/Cornell_University#Financial_aid.
78. www.duke.edu.
79. https://facts.duke.edu/.
80. https://en.wikipedia.org/wiki/Duke_University.
81. https://en.wikipedia.org/wiki/University_of_Cambridge.
82. https://www.cam.ac.uk/research/news/machine-learning-algorithm-helps-in-the-search-for-new-drugs.
83. 3https://www.cam.ac.uk/about-the-university?ucam-ref=home-quicklinks.
84. www.cam.ac.uk
85. https://en.wikipedia.org/wiki/Johns_Hopkins_University.
86. https://www.reuters.com/innovative-universities-2018/profile?uid=19.
87. https://www.hopkinsmedicine.org/the_johns_hopkins_hospital/about/enhanced_facilities/facts_figures.html.
88. https:// www.u-tokyo.ac.jp.
89. https://www.u-tokyo.ac.jp/en/about/about.html.
90. https://en.wikipedia.org/wiki/University_of_Tokyo.
91. https://www.reuters.com/innovative-universities-2018/profile?uid=31.
92. https://en.wikipedia.org/wiki/University_of_Erlangen%E2%80%93Nuremberg.
93. www.utoronto.ca.
94. https://www.utoronto.ca/about-u-of-t/quick-facts.
95. https://www.utoronto.ca/about-u-of-t/reports-and-accountability.
96. https://en.wikipedia.org/wiki/University_of_Toronto.
97. https://www.reuters.com/innovative-universities-2018/profile?uid=44.
98. https://en.wikipedia.org/wiki/Tsinghua_University.
99. https://www.reuters.com/innovative-universities-2018/profile?uid=55.
100. www.umontpellier.fr.
101. https://en.wikipedia.org/wiki/University_of_Montpellier.
102. https://www.umontpellier.fr/university-of-montpellier.
103. https://www.dtu.dk.
104. https://www.dtu.dk/english/About/FACTS-AND-FIGURES.
105. https://en.wikipedia.org/wiki/Technical_University_of_Denmark.
106. https://www.tudelft.nl.
107. https://www.tudelft.nl/en/about-tu-delft/facts-and-figures.
108. https://en.wikipedia.org/wiki/Delft_University_of_Technology.
109. http:// www.nus.edu.sg.
110. https://www.nus.edu.sg/about#corporate-information.
111. https://en.wikipedia.org/wiki/National_University_of_Singapore#Alumni.
112. Zahed, A. H. M. et al, The Way to Produce Pioneering Research at King Abdulaziz University, Final Report of Project No. MBD/29/015, Jeddah 2016. (In Arabic).
113. 113. Gill, I. and H. Kharas, An East Asian Renaissance: Ideas for Economic Growth, Washington D.C.: World Bank, 2007.
114. Im, F. G. and Rosenblatt D., Middle-Income Traps, A Conceptual and Empirical Survey, Policy Research Working Paper, World Bank, 2013.
115. Zahed, A. H. M. et al, Excellent Practices in Scientific Research of the Pioneering Universities, Final Report of Project No. MBD/23/016, Jeddah 2017. (In Arabic).
116. William G. Tierney, "The Creative University", Private communication.
117. Vision 2030 of the Kingdom of Saudi Arabia, https://www.vision2030.gov.sa/en/node.